Proceedings of the International Colloquium in Textile Engineering, Fashion, Apparel and Design 2014 (ICTEFAD 2014)

Mohd Rozi Ahmad • Mohamad Faizul Yahya
Editors

Proceedings of the International Colloquium in Textile Engineering, Fashion, Apparel and Design 2014 (ICTEFAD 2014)

Springer

Editors
Mohd Rozi Ahmad
Mohamad Faizul Yahya
Faculty of Applied Sciences
Universiti Teknologi MARA
Shah Alam
Malaysia

ISBN 978-981-10-1362-1 ISBN 978-981-287-011-7 (eBook)
DOI 10.1007/978-981-287-011-7
Springer Singapore Heidelberg New York Dordrecht London

Printed on acid-free paper

Springer is part of Springer Science+Business Media (www.springer.com)

Contents

Production of Bicomponent Polymer Droplets by Electrospraying

A. Jadhav, L. Wang, and R. Padhye

Abstract

Electrospraying is a method of generating a fine mist through electrostatic charging. It is used for producing polymer droplets at submicron range, which enhances functional properties of the substrate after deposition by providing thin level coating and larger surface area. In this research, in order to impart multifunctionality to substrate, the two distinct polymers thermoplastic polyurethane (TPU) and poly vinyl chloride (PVC) are electrosprayed simultaneously by specially designed nozzle. Energy dispersive spectroscopy (EDS) was used to confirm the bicomponent droplet fabrication. The results show that new electrospraying system demonstrated the feasibility of producing bicomponent TPU/PVC polymer droplets.

Keywords

Electrospraying • Bicomponent droplets • Thin-film deposition • Thermoplastic polyurethane • Polyvinyl chloride

Introduction

Electrospraying is a method of generating a fine mist through electrostatic charging. As the liquid passes through a nozzle, fine droplets are generated by electrically charging the liquid to a very high voltage [1]. Charged droplets are self-dispersing in space, resulting in the absence of droplet coagulation. The deposition efficiency of a charged spray on an object is higher than for an uncharged spray. This feature can be advantageous, for example, in surface coating, thin-film production and electroscrubbing. Electrospraying can be widely applied to both industrial processes and scientific instrumentation [2]. Electrospraying is used for micro- and nano-thin-film deposition [3], micro- or nanoparticle production and micro- or nano-capsule formation. Thin films and fine powders are (or potentially could be) used in modern material technologies, microelectronics and medical technology. The effects of several process parameters, such as the applied voltage, electric field strength, flow rate, concentration and distance between the nozzle and collector, have been explored in great detail for thermoplastic polyurethane (TPU) [4–6].

Many researchers have reported the feasibility of bicomponent electrospinning systems. Blends of polyaniline, a conducting polymer, with polyethylene oxide (PEO) dissolved in chloroform were electrospun to produce filaments in the range of 5–30 nm [7, 8]. Regenerated silk was blended with PEO to eliminate the expansion of brittle β-sheets of silk fibroin [9]. Hence, utility of the resulting electrospun mats was improved in vitro and in vivo conditions. There have been no reported studies regarding bicomponent droplet fabrication by electrospraying. Hence, an attempt has been made to design and develop bicomponent electrospraying system. Some unique characteristics must be considered when electrospraying was performed from blends of polymer solutions. For a blend of two polymers (in the same solvent or different

This research work is supported by DIISRTE, Australian Govt., in the form of Endeavour scholarship to the first author.

A. Jadhav (✉) • L. Wang • R. Padhye
School of Fashion and Textiles, RMIT University, Melbourne, VIC, Australia
e-mail: amit.jadhav@rmit.edu.au

M.R. Ahmad and M.F. Yahya (eds.), *Proceedings of the International Colloquium in Textile Engineering, Fashion, Apparel and Design 2014 (ICTEFAD 2014)*, DOI 10.1007/978-981-287-011-7_1,

solvents), the mixture should be homogenous so that the resultant droplets possess a uniform spatial configuration. In addition to being thermodynamically miscible, the interactions between the polymer and the solvent of the opposing pair are of critical importance when two polymers are dissolved in two solvents respectively. Hence, during electrospraying of two polymers, the thermodynamic and kinetic aspects of solution need to be considered. Another way to produce electrosprayed droplets from two polymers is by electrospraying two polymers simultaneously in a side-by-side arrangement.

In this research, the developed bicomponent system utilised was such that the two polymer solutions do not come into physical contact until they reached the end of the nozzle, where the process of droplet formation began. The electrospraying device was designed in such way that two polymer solutions were electrosprayed simultaneously in a side-by-side manner. This allows for a bicomponent electrospray coating that possesses properties from each of the polymeric components. For instance, one of the polymers could contribute to mechanical strength, while the other could impart hydrophobic properties to the resulting textile substrate. This could be useful for a protective textile application. In fact by suitably choosing the constituent components based on their respective properties, the potential of these bicomponent droplets to be utilised in various applications becomes enhanced. These applications could include medical, protective and technical textiles. The prime objective of this study was to demonstrate the feasibility of this new methodology to produce bicomponent droplets via electrospraying at submicron range. In this study, the new bicomponent electrospraying device is described. Preliminary results on TPU/polyvinyl chloride (PVC) bicomponent droplets are presented. It is important to note here that TPU/PVC is a miscible system. PVC has a glass transition temperature of 85 °C and is therefore a glassy and stiff material at room temperature. The mechanical properties of PVC, especially toughness, can be enhanced by suitable plasticisation [10]. When blended with a thermoplastic urethane polymer, it was expected that the resulting mechanical toughness and hydrophobicity would be improved without sacrificing original property, depending on the composition ratio. It was expected that the resultant coating on the substrate will possess characteristics of both the components. The main reason behind using such a technique is to enable easy identification of the two components in each polymer pair by energy dispersive spectroscopy (EDS), thereby demonstrating the feasibility of electrospraying a bicomponent droplet coating. The EDS detector, which was a part of the scanning electron microscope (SEM), was utilised in this study to investigate the morphology of the droplets. This allowed characterisation of the local composition of the bicomponent droplets at the submicron level.

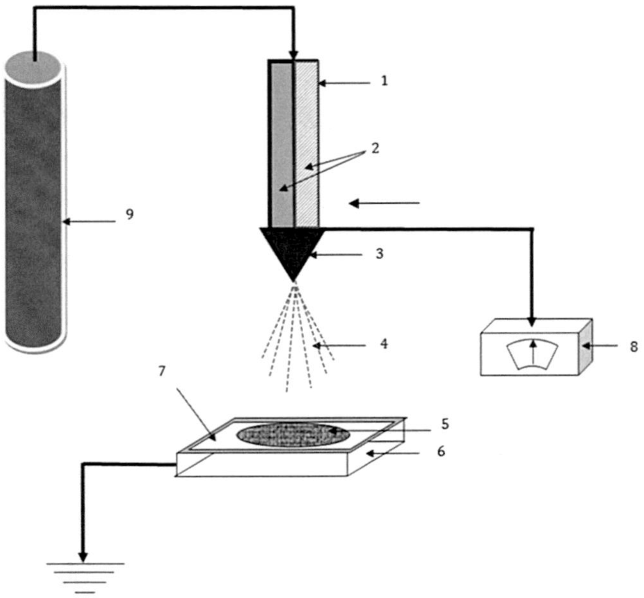

Fig. 1 Equipment setup for bicomponent electrospraying

Experimental

Chemicals

PVC of M_w of 75,400 g/mol in the form of a fine powder and TPU of M_w 30,000 in the form of chips received. The THF was used as solvent for dissolving TPU. N, N-Dimethylacetamide was used to dissolve PVC at different weight concentrations ranging from 1 to 5 % w/w. All chemicals were used as received without any purification.

Electrospraying

Schematic arrangement of the bicomponent electrospraying instrument parts is shown in Fig. 1.

The electrospraying apparatus consisted of a solution container (1) which was filled with the required polymer solutions (2). The container has a capacity of 150 ml and a specially designed nozzle (3) that was connected to the positive terminal. The power source (8) was capable of providing output voltage from 0 to 30 kV. The inner diameter of nozzle (3) was 0.5 mm. A metal plate was used as a collector (6) on which the fabric substrate (5) was placed to be coated with the polymeric solution. Compressed nitrogen gas (9) was used to apply pressure inside the solution container in order to push the polymer solution towards the tip of the nozzle and to maintain a constant flow rate. The pressure was between the range of 0 and 16 bar.

SEM

FEI Quanta Nova 200 scanning electron microscopy (SEM) was used to study the TPU/PVC droplet morphology.

EDS

EDS was performed on TPU/PVC electrosprayed substrates to calibrate the peak intensities of oxygen (O) and chlorine (Cl) with the actual composition in the blend. Energy dispersive X-ray spectroscopy (EDS) is a qualitative and quantitative X-ray microanalytical technique that can provide information on the chemical composition of a sample for elements with atomic number (Z) >3.

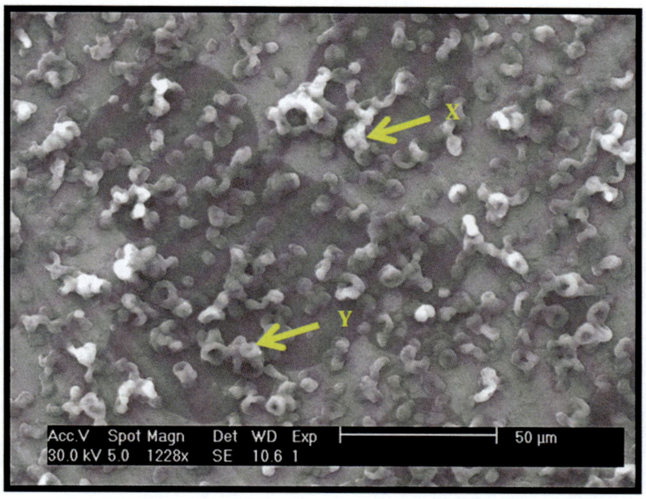

Fig. 2 SEM of bicomponent PVC (X) and TPU (Y) droplet deposition

Results and Discussions

As shown in Fig. 1, the two polymer solutions came into contact only at the tip of the nozzle. Even though the two polymer solutions are charged to the same polarity, some amount of mixing of the two components was expected to take place as the two solutions reached the end of the nozzle tip. Under stable electrospraying conditions, a fluctuating jet was observed for TPU/PVC at 25 kV, a target distance of 20 cm and total flow rate of 3 ml/h at 10 bar pressure on solution inside the cylinder. Interestingly, when the nozzle–collector distance was 10 cm or larger, a single common Taylor cone was observed. From the surface of this Taylor cone, a fluctuating jet was ejected. The position of ejection of the jet on the surface of the Taylor cone changed very rapidly with time and led to a somewhat nonsteady flow of the polymer solution. These fluctuations likely influence the extent of mixing of the two charged solutions when they came into contact at the tip of the nozzle.

At distances larger than 20 cm, the jet was not continuous and the solutions were dripped due to weak field strength that did not convey the jet to the collector. At distances less than 10 cm, two Taylor cones were observed to emanate from each of the two holes of nozzle. As a result, two jets were observed to eject from each Taylor cone under these conditions. At such low distances (<10 cm), the field strength was relatively strong, thereby inducing a strong electrostatic repulsion between the two polymer solutions emanating from each hole of nozzle.

This leads to the formation of two Taylor cones and subsequently two separate but identically charged (in terms of polarity) jets causing the formation of two zones of droplet collection on the target, each corresponding to only one of the two respective polymer components. Hence, bicomponent electrospraying was conducted at nozzle–collector distances of 10–20 cm in this study.

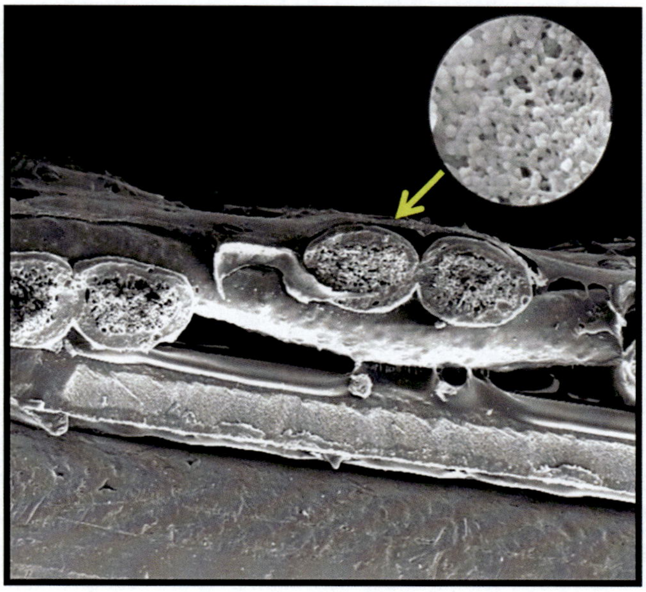

Fig. 3 Bicomponent droplet formation on textile substrate

An SEM image of the dried TPU/PVC electrosprayed droplets can be seen in Fig. 2. It can also be seen that the PVC droplets appear to be brighter than the TPU polymer droplets. Figure 3 shows the bicomponent coated film formation on the textile substrate. The EDS was performed on several spatial positions within the sprayed area, but of particular importance are the two regions marked as X and Y. In Fig. 2, region X exhibited an intense peak of chlorine (Cl), indicating the local composition to be principally comprised of PVC (Fig. 4), whereas region Y exhibited an intense peak of oxygen (O), indicating the local dominance of TPU (Fig. 5). It is important to note the presence of a

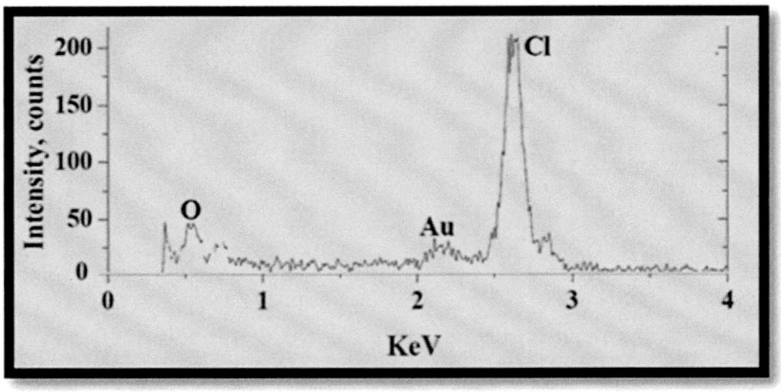

Fig. 4 EDS of region X

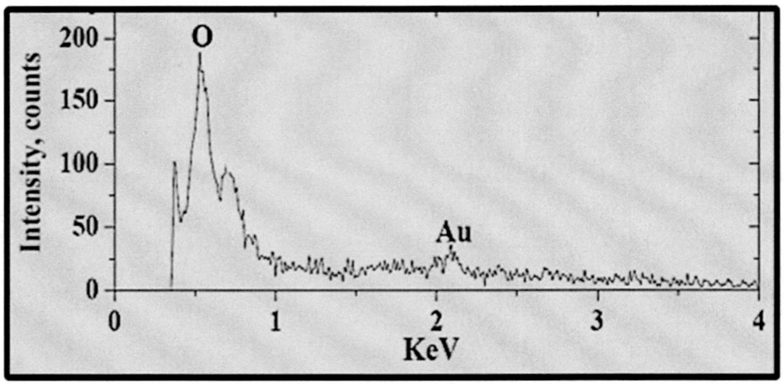

Fig. 5 EDS of region Y

smaller peak corresponding to oxygen in Fig. 4. This peak was relatively weak but it does indicate the presence of TPU in the predominantly PVC-rich polymer droplet at region X. From the EDS mapping, it can be concluded that although droplets, which composed primarily of either component, were observed to form, the presence of trace amounts of the other component indicates some level of physical mixing of the two solutions. These variations in the composition of the bicomponent droplets are attributed to the fluctuations of the common jet on the surface of the Taylor cone. Chain diffusion and relaxation could also affect the mixing of the two components. It is useful to consider that both the jets emanating from the nozzle tip carry the same electrostatic charge and therefore experience mutual electrostatic explosion.

In addition, PVC can develop a low degree of crystallinity when solidified from solution. Hence, the amount of crystallinity induced in these droplets as the solvent evaporated during the flight of the droplets to the grounded target would also influence the extent of mixing of the two components.

Therefore, the mixing of the two charged solutions in bicomponent electrospraying is a competing phenomenon between the several effects of the fluctuations of the jet, electrostatic repulsion or explosion between the similarly charged jets, diffusion of polymeric chains of one component in the other, chain relaxation and evaporation rate of the solvent and solvent-induced crystallisation in semi-crystalline polymers.

Conclusion

Two polymer solutions were electrosprayed simultaneously in a side-by-side arrangement to produce bicomponent droplets of TPU/PVC, and EDS was utilised to identify the respective components by detecting the signal corresponding to chlorine and oxygen from TPU/PVC droplets. It was found that the two types of polymer droplets are deposited on the target substrate. Utilising the methodology described in this study, the feasibility of electrospraying of bicomponent droplets has been demonstrated.

Acknowledgement The authors acknowledge the facilities, and the scientific and technical assistance, of Australian Microscopy & Microanalysis Research facility at RMIT Microscopy & Microanalysis Facility, at RMIT University.

References

1. A. Jaworek, A. Krupa, Jet and drops formation in electrohydrodynamic spraying of liquids. A systematic approach. Exp. Fluids **27**, 43–52 (1999)
2. O.V. Salata, Tools of nanotechnology: electrospray. Curr. Nanosci. **1**, 25–33 (2005)

3. N. Arya, S. Chakraborty, N. Dube, D.S. Katti, Electrospraying: a facile technique for synthesis of chitosan-based micro/nanospheres for drug delivery applications. J. Biomed. Mater. Res. B Appl. Biomater. **88**, 17–31 (2009)
4. A. Jadhav, L. Wang, R. Padhye, Influence of applied voltage on droplet size distribution in electrospraying of thermoplastic polyurethane. Int. J. Appl. Mech. Mater. **2**, 64 (2013)
5. A. Jadhav, L.J. Wang, R. Padhye, Effect of applied voltage on polymer aggregation in electrospraying of thermoplastic polymer. Adv. Mater. Res. **535**, 2522–2525 (2012)
6. A. Jadhav, L. Wang, R. Padhye, Effect of field strength on polymer aggregation in electrohydrodynamic spraying of thermoplastic polyurethane. Appl. Mech. Mater. **328**, 895–900 (2013)
7. N. Pinto, A. Johnson, A. MacDiarmid, C. Mueller, N. Theofylaktos, D. Robinson et al., Electrospun polyaniline/polyethylene oxide nanofiber field-effect transistor. Appl. Phys. Lett. **83**, 4244–4246 (2003)
8. P. Kahol, N. Pinto, An EPR investigation of electrospun polyaniline-polyethylene oxide blends. Synth. Met. **140**, 269–272 (2004)
9. H.-J. Jin, S.V. Fridrikh, G.C. Rutledge, D.L. Kaplan, Electrospinning *Bombyx mori* silk with poly (ethylene oxide). Biomacromolecules **3**, 1233–1239 (2002)
10. J. Cano, M. Marιn, A. Sanchez, V. Hernandis, Determination of adipate plasticizers in poly (vinyl chloride) by microwave-assisted extraction. J. Chromatogr. A **963**, 401–409 (2002)

Characteristics of Electrospun PVA-*Aloe vera* Nanofibres Produced via Electrospinning

N.A. Abdullah@Shukry, K. Ahmad Sekak, M.R. Ahmad, and T.J. Bustami Effendi

Abstract

The aim of this work was to investigate the characteristics of polyvinyl alcohol (PVA) nanofibres with the incorporation of *Aloe vera* as a novel polymer-drug carrier. The nanofibres were produced via the electrospinning technique. The morphological structure and properties were characterized using field emission scanning electron microscopy (FESEM), Fourier transform infrared spectroscopy (FTIR) and differential scanning calorimetry (DSC). The FESEM image shows homogenous and linear fibre when PVA is mixed with *Aloe vera*. The average size was 123 nm, smaller than the PVA nanofibre. The presence of aloin in DSC showed that the *Aloe vera* was successfully embedded within the PVA nanofibre.

Keywords

Nanofibres • Electrospinning • PVA • *Aloe vera*

Introduction

Electrospinning is a very simple yet versatile method of creating polymer-based high-functional and high-performance nanofibres where it can revolutionize the world of structure materials including the textile field, wound dressings [1, 2], drug delivery systems and tissue engineering [2] in biomedical site or as a membrane of filtration [3] and electronic component coating in an industrial application. Electrospinning, as the name implies, is a process by which nanofibres from a solution or melt (polymer or polymer mixed) are generated in the presence of electric field. Unlike other techniques of producing 1D polymeric nanostructure like top-down method which involves photolithography or soft lithography, electrospinning is incredibly straightforward, simple and cheaper. Instead of using mechanical force of pulling or stretching like conventional spinning process, electrospinning used electrostatic or coulombic [4–7] force as the driving mechanism in drawing nanofibres (Fig. 1). But the most remarkable effect of electrospinning is that it can draw fibres with a diameter as low as tens of nanometres.

Electrospun nanofibre web exhibits a number of outstanding properties such as high surface area compared to film, light weight [1, 8–12] and high porosity [2]. Through the interesting characteristics of nanofibres, many researches were done on various types of polymer as the carrier in drug delivery application [13–17]. PVA has been identified as the suitable candidate that can be used in releasing biological and medical materials in a controlled way [13]. PVA has unique properties such as good chemical resistance, thermal stability, biocompatibility and nontoxicity which make its suitable to be electrospun as polymer-drug carrier [18]. PVA has also been studied intensively due to its good film forming and physical properties, high hydrophilicity and processability [18]. In the past few years, research on electrospun PVA nanofibres has looked at various parameters such as solution concentration, solution flow rate,

N.A. Abdullah@Shukry (✉) • K. Ahmad Sekak • M.R. Ahmad
Faculty of Applied Sciences, Universiti Teknologi MARA, 40450 Shah Alam, Selangor, Malaysia
e-mail: n.athirahabdullah@yahoo.com

T.J. Bustami Effendi
Faculty of Pharmacy, Universiti Teknologi MARA, 40450 Shah Alam, Selangor, Malaysia

M.R. Ahmad and M.F. Yahya (eds.), *Proceedings of the International Colloquium in Textile Engineering, Fashion, Apparel and Design 2014 (ICTEFAD 2014)*, DOI 10.1007/978-981-287-011-7_2,

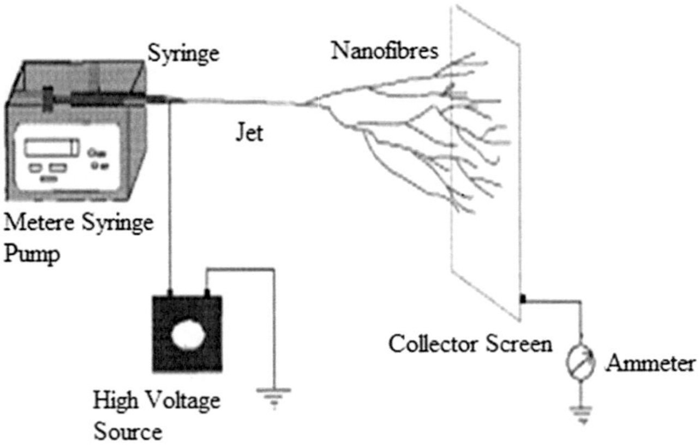

Fig. 1 Electrospinning process

degree of hydrolysis, applied voltage, tip-target distance, ionic salt addition, pH and surfactant addition that affect the morphological structures and diameters of electrospun PVA [1, 3, 18]. Therefore, it has become more interesting to explore *Aloe vera* as a drug carrier.

The aim of this work was to produce PVA-*Aloe vera* nanofibres via electrospinning. *Aloe vera* is known as the oldest therapeutic herb and has the ability to promote wound healing as well as treat burn area on the skin. It consists of two important parts in which the outer layer is called vascular bundle and the inner layer is known as colourless parenchyma containing *Aloe vera* gel. *Aloe vera* has three compositions that include structural, chemical and polysaccharide. Polysaccharide plays the major role in promoting wound healing. It contains salicylic acid together with other chemical components such as proteins, lipids, amino acids, vitamins, enzymes, inorganic compounds and small organic compounds [19–21]. Not many research works are done on the incorporation of *Aloe vera* with polymer membrane. Therefore, this study attempts to encapsulate *Aloe vera* drugs in electrospun polymeric nanofibres through electrospinning. This was done by dissolving PVA mixed with *Aloe vera* in deionized water. The solution was electrospun to form long and continuous nanofibre. For comparison, PVA nanofibre membrane was prepared. The morphological structures of PVA-*Aloe vera* nanofibres were characterized using field emission scanning electron microscopy (FESEM) and different scanning calorimetry (DSC), and the functional group was determined using Fourier transform infrared spectroscopy (FTIR).

Materials and Methods

Materials

Polyvinyl alcohol (PVA) (Mw: 125,000 g/mol) was purchased from Sigma-Aldrich. *Aloe vera* powder extract was received and used without further purification. Distilled water was used as the solvent in the experiment.

Sample Preparation

10 g of PVA powder was dissolved into 90 ml of distilled water at 10 % w/v concentration. The solution was stirred at 80 °C for 3 h using electromagnetic stirrer to get homogenous and crystal clear solution.

Meanwhile, 5 % of *Aloe vera* powder extract (according to 10 g of PVA) was mixed with PVA at 10%v/w and stirred directly using the same method as mentioned above. Both solutions were cooled down at room temperature for several hours before electrospinning.

Electrospinning

Electrospinning was carried by setting the positive electrode on high-voltage DC power supply to the solution that contained 3 ml syringe. The negative electrode was connected to the needle that was used as the nozzle and the grounded electrode to a rotating metal drum wrapped with aluminium foil. The solutions were carefully loaded inside the syringe during the electrospinning process. A pressure was applied on top of the syringe to maintain a steady flow of polymer solution from capillary needle. The tip of the needle and ground collector were placed horizontally facing each other (Fig. 1).

The voltage was applied at 15 kV across the distance of 8 cm between the tip of the needle and aluminium collector. The feed rate was controlled at 0.5 mol/h and the solution was electrospun for about 6 h in order to produce a neat polymer-drug membrane. A collector was rotated at 50 rpm. Finally, the nanofibres were removed from the collector and placed in the oven overnight at 37 °C.

Membrane Characterization

The morphological structure and fibre diameter of electrospun PVA and PVA-AV nanofibre were observed under FESEM (Zeiss) at 2 kV. The membranes were coated with platinum for 3 min. Three hundred readings of fibre diameter were collected and measured with the image visualization software, ImageJ.

The functional groups of the sample were characterized using FTIR (Perkin Elmer).

Differential Scanning Calorimetry (DSC)

The prepared samples were measured at glass transition over a temperature range about 30–120 °C and possibly a thermal degradation range from 250 to 350 °C with 10 °C/min under DSC (Perkin Elmer).

Fig. 2 10 % w/v PVA nanofibre

Fig. 3 10 % w/v of PVA, 5 % w/v of AV

Result and Discussion

Morphological Structure of Electrospun PVA and PVA-AV Nanofibre

Figure 2 shows the image of fine fibre formation and homogeneous and continuous fibre for pure PVA nanofibres. There is no formation of beads within the PVA nanofibres. The average diameter of PVA nanofibres was 168 nm.

Meanwhile, Fig. 3 shows the image of PVA-AV nanofibres in fine, homogenous and linear fibres. There is also no presence of beads indicating that the *Aloe vera* powder had been completely entrapped inside the PVA membrane. This was also discussed by Li et al. [22] where the riboflavin and caffeine were encapsulated homogeneously in the PVA matrices. However, the fibres are slightly distorted but there is no formation of branch in PVA-AV nanofibres.

The average diameter of PVA-AV nanofibre was reduced to 123 nm as compared with PVA nanofibres. This may be because of the interaction between the hydrogen groups in PVA and *Aloe vera* which increased the coulombic repulsion and electrostatic force where the molecules pull closer to each other. It was explained by Taepaboon et al. [16] that coulombic repulsion works to stretch the charged jet and electrostatic force brings the jet to the target. It also involves the conductivity of the solution. Due to the presence of *Aloe vera*, the conductivity may have decreased which leads to beaded-free fibre formation.

Fourier Transform Infrared Spectroscopy

The infrared spectra (Fig. 4) illustrate the characteristic of PVA and PVA-AV. The peak appeared at 3,480 cm^{-1} in *Aloe vera* extract powder, which indicates the presence of phenolic-OH group. *Aloe vera* consists a lot of chemical compound and phenolic-OH group is one of them where it can be found in anthraquinones. Similar results also were achieved by [23, 24].

The peak that also appeared at 3,400–3,380 cm^{-1} in both PVA and PVA-AV indicates the presence of hydroxyl group in the membrane. It is due to the properties of PVA and *Aloe vera* that both contain hydroxyl group. The peak (b) at 2,780 cm^{-1}, however, was slowly distinguished when PVA

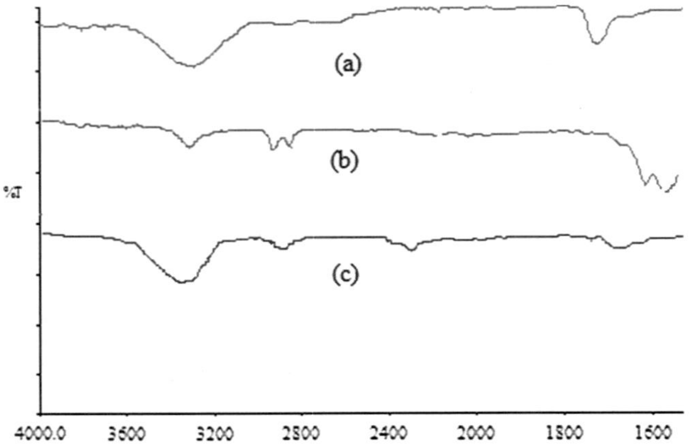

Fig. 4 FTIR spectra for (**a**) *Aloe vera* extract powder, electrospun fibres of (**b**) 10 %w/v of PVA and (**c**) 10 % w/v of PVA, 5 % w/v of AV

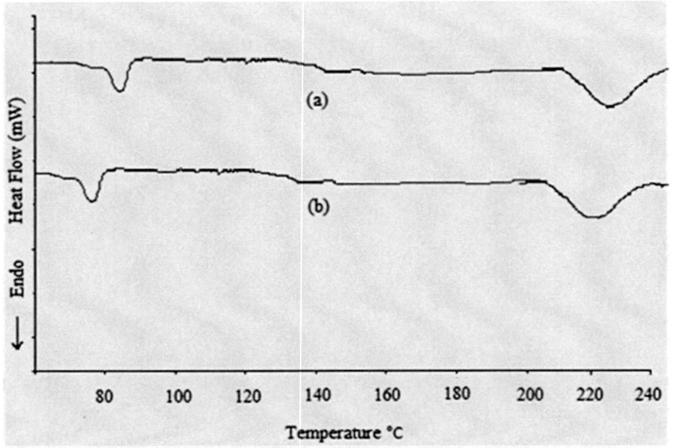

Fig. 5 DSC thermograms for nanofibre of (**a**) 10 % w/v PVA and (**b**) 10 % w/v PVA and 5 % of AV

was mixed with *Aloe vera* due to the stretching of hydroxyl group that contained in both polymer and drug. It also attributed to intermolecular interaction between the hydrogen group with carbonyl among PVA chains and *Aloe vera* due to hydrophilic forces when the peak at Fig. 4c shifts from 1,679 to 1,690 cm^{-1}.

Differential Scanning Calorimetry (DSC)

The DSC thermograms for PVA and PVA-AV nanofibre membranes are shown in Fig. 5. The peak of glass transition occurred at 81 °C, indicating pure PVA nanofibre. However, the peak reduced to 76 °C in glass transition when adding the *Aloe vera* powder. This is due to the presence of aloin in *Aloe vera* [23, 24]. Aloin can be found in polysaccharide that contains anthraquinones. It is particularly used for medical purposes such as for skin treatment and beauty cream product. At melting point, when (b) shifted to 222 °C from 226 °C, it may also be due to the presence of aloin. This phenomenon is due to small particles of *Aloe vera* extract powder that completely dissolved in PVA matrix which affect the crystal structure as well as form an amorphous structure. However, the results contradicted with Ulsu et al. [15] where the peak was reduced at melting point. It may be because of the state of *Aloe vera* in extract powder form that gave better result as compared to *Aloe vera* in a gel form. It can be said that the *Aloe vera* extract powder was entrapped inside PVA which can provide promising wound dressing.

Conclusion

The fabrication of electrospun nanofibres using *Aloe vera* was successfully done and homogenously mixed with PVA. The SEM images show reduction of diameter in electrospun PVA-AV nanofibre as compared to electrospun PVA membrane. Besides, the functional group also indicates that there is no formation of new groups in the membrane proving that both PVA and *Aloe vera* were compatible. The shift of DSC thermograms occurred at melting point provided solid evidence on the effect of *Aloe vera* in electrospun PVA membrane.

Acknowledgement The authors would like to acknowledge the grant provided by the Ministry of Education through the Exploratory Research Grant Scheme (ERGS). The assistance from the Research Management Institute (RMI) of Universiti Teknologi MARA is greatly appreciated.

References

1. T. Subbiah, G.S. Bhat, R.W. Tock, S. Parameswaran, S. Ramkumar, Electrospinning of nanofibers. J. Appl. Polym. Sci. **96**(2), 557–569 (2005)
2. R. Barhate, C. Loong, S. Ramakrishna, Preparation and characterization of nanofibrous filtering media. J. Membr. Sci. **283**, 209–218 (2006)
3. A. Frenot, I.S. Chronakis, Polymer nanofibers assembled by electrospinning. Curr. Opin. Colloid Interface Sci. **8**(1), 64–75 (2003)
4. S. Tan, X. Huang, B. Wu, Some fascinating phenomena in electrospinning processes and applications of electrospun nanofibers. Polym. Int. **56**(11), 1330–1339 (2007)
5. C. Feng, K. Khulbe, T. Matsuura, Recent progress in the preparation, characterization, and applications of nanofibers and nanofiber membranes via electrospinning/interfacial polymerization. J. Appl. Polym. Sci. **115**(2), 756–776 (2010)
6. P. Lu, B. Ding, Applications of electrospun fibers. Recent Patents Nanotechnol. **2**(3), 169–182 (2008)
7. S. Ramakrishna, K. Fujihara, W. Teo, T. Lim, Z. Ma, *An Introduction to Electrospinning and Nanofibers* (World Scientific, Singapore, 2005)
8. A.P.S. Sawhney, B. Condon, K.V. Singh, S.S. Pang, G. Li, H. David, Modern applications of nanotechnology in textile. Text. Res. J. **78**(8), 731 (2008)
9. D.R. Salem, *Structure Formation in Polymeric Fibers* (Hanser Publishers, Munich, 2001)

10. O.O. Dosunmu, G.G. Chase, J. Varabhas, W. Kataphinan, D. Reneker, Polymer nanofibers from multiple jets produced on a porous surface by electrospinning. Nanotechnology **17**(4), 1123–1127 (2006)
11. N. Affandi, Y. Truong, I. Kyratzis, R. Padhye, L. Arnold, A non-destructive method for thickness measurement of thin electrospun membranes using white light profilometry. J. Mater. Sci. **45**, 1411–1418 (2010)
12. N.D. Nor Affandi, M.R. Ahmad, A. Baharudin, N.A. Abdullah Shukry, Effect of crosslinking on the solubility and morphological structures of the PVA nanofibres. Paper presented at the humanities, IEEE colloquium on science and engineering (CHUSER) (2012)
13. E.R. Kenawy, F.I. Abdel-Hay, M.H. El-Newehy, G.E. Wnek, Controlled release of ketoprofen from electrospun poly (vinyl alcohol) nanofibers. Mater. Sci. Eng. A **459**, 390–396 (2007)
14. J. Zeng, L. Yang, Q. Liang, X. Zhang, H. Guan, X. Xu, X. Jing, Influence of the drug compatibility with polymer solution on the release kinetics of electrospun fiber formulation. J. Control. Release **105**(1), 43–51 (2005)
15. I. Uslu, S. Keskin, A. Gül, T.C. Karabulut, M.L. Aksu, Preparation and properties of electrospun poly (vinyl alcohol) blended hybrid polymer with *Aloe vera* and HPMC as wound dressing. Hacettepe J. Biol. Chem. **38**, 19–25 (2010)
16. P. Taepaiboon, U. Rungsardthong, P. Supaphol, Drug-loaded electrospun mats of poly (vinyl alcohol) fibres and their release characteristics of four model drugs. Nanotechnology **17**(9), 2317 (2006)
17. N. Bölgen, Y.Z. Menceloğlu, K. Acatay, I. Vargel, E. Pişkin, In vitro and in vivo degradation of non-woven materials made of poly (ε-caprolactone) nanofibers prepared by electrospinning under different conditions. J. Biomater. Sci. Polym. Ed. **16**(12), 1537–1555 (2005)
18. E. Yang, X. Qin, S. Wang, Electrospun crosslinked polyvinyl alcohol membrane. Mater. Lett. **62**(20), 3555–3557 (2008)
19. J.H. Hamman, Composition and applications of *Aloe vera* leaf gel. Molecules **13**(8), 1599–1616 (2008)
20. R. Haniadka, P. Kamble, A. Azmidha, P.P. Mane, Geevarughese, P. L. Palatty, M.S. Baliga, in *Review on the Use of Aloe vera (Aloe) in Dermatology*. Bioactive Dietary Factors and Plant Extracts in Dermatology (Springer, New York, 2013), pp. 125–133
21. X. Li, M.A. Kanjwal, L. Lin, I.S. Chronakis, Electrospun polyvinyl-alcohol nanofibers as oral fast-dissolving delivery system of caffeine and riboflavin. Colloids Surf. B Biointerfaces **103**, 182 (2012)
22. S. Ravi, P. Kabilar, S. Velmurugan, R.A. Kumar, M. Gayathiri, Spectroscopy studies on the status of aloin in *Aloe vera* and commercial samples. J. Exp. Sci. **2**(8), 10–13 (2011)
23. A.J. Amalraj, J.W. Sahayaraj, C. Kumar, S. Rajendran, A.P.P. Regis, S.K. Selvaraj, R. Mohan, Corrosion inhibitor *Aloe vera* – Nickel system controlling the corrosion of carbon steel in rain water, 315–319 (2013)
24. C.I. Nindo, J.R. Powers, J. Tang, Thermal properties of *Aloe vera* powder and rheology of reconstituted gels. Trans. ASABE **53**(4), 1193–120 (2010)

Puncture Strength of Natural Rubber Latex Coated Unidirectional Fabric After Heat Ageing

M.R. Ahmad, A.L. Anisah, N.V. David, and W.Y.W. Ahmad

Abstract

This article presents the effect of ageing on the puncture strength of natural rubber latex (NRL) coated ultra high molecular weight polyethylene (UHMWPE) unidirectional (UD) fabrics. Three different coating depths, namely single dip, double dip and triple dip, were used. The samples were heated at 40, 70 and 100 °C for 48 and 192 h. The samples were also placed inside a car for 2–8 months duration. In general, the puncture strength of the samples decreases as the heat ageing temperature increases. However, at 70 °C, the strength increases with extended time. Changes in the structure of the materials tested were estimated with Fourier transform infrared spectroscopy (FTIR).

Keywords

Puncture strength • Heat ageing • Storage • Natural rubber latex • Ultra high molecular weight polyethylene • Unidirectional • FTIR

Introduction

Ultra high molecular weight polyethylene (UHMWPE) unidirectional (UD) fabric has superior properties in terms of strength and low specific weight (0.97 g/cm^3) which are advantageous for protective clothing. Karahan et al. [1] compared the ballistic performance and energy absorption capabilities of woven and the UD fabric structures and found that the UD fabric has lower trauma depth and higher trauma diameter. This means that the UD fabric absorbed more energy than woven fabrics. Another study [2] shows that UD fabric allows energy to dissipate along the fibres much faster than in conventional woven fabric.

Some studies showed that UHMWPE coated with natural rubber latex (NRL) enhances the absorption characteristics of the material [3]. This is supported by Hassim et al. [2] which found that the puncture strength of UHMWPE coated NRL increases because the fibres are trapped within the NRL layer. However, ageing will take place and will reduce the absorption properties of the material. Therefore, an investigation on the thermal and heat storage capabilities of the UD fabric is necessary. The results obtained after heat ageing studies will provide useful information on the mechanical performance of the fabric.

UHMWPE is a type of polymer with a semicrystalline structure which means that it exhibits both an amorphous and crystalline region. The polymer aged rapidly under the effect of light, oxygen and heat leading to loss of strength, stiffness and flexibility [4]. Heat can cause the formation of free radicals that can react together to form a cross-link between the polymer chain and improve the wear resistance of the UHMWPE [5]. On the other hand, reaction of the free radicals with oxygen present in the material can cause oxidative chain scission, shortening of the molecular chains which degrade the wear behaviour of the UHMWPE [6].

Kennedy et al. [7] showed that temperature above 63 °C can cause unirradiated UHMWPE to have subsurface oxidation similar to that caused by shelf ageing of irradiated

M.R. Ahmad • A.L. Anisah (✉) • W.Y.W. Ahmad
Faculty of Applied Sciences, Universiti Teknologi MARA, 40450 Shah Alam, Selangor Darul Ehsan, Malaysia
e-mail: anisah0307@gmail.com

N.V. David
Faculty of Mechanical Engineering, Universiti Teknologi MARA, 40450 Shah Alam, Selangor Darul Ehsan, Malaysia

M.R. Ahmad and M.F. Yahya (eds.), *Proceedings of the International Colloquium in Textile Engineering, Fashion, Apparel and Design 2014 (ICTEFAD 2014)*, DOI 10.1007/978-981-287-011-7_3,

UHMWPE. Systematic studies concerning UHMWPE have been published by several authors [4–12] which found that heat ageing will alter the macromolecular structure of the material and results in the reduction of the material's performance. Other studies [13–16] also showed that environment and humidity cause a remarkable reduction in the mechanical properties of the materials.

Nevertheless, NRL, which acts as the coating material, may also give a significant effect. Agosti et al. [17] proposed that degradation process for latex is caused by heating. Different temperatures result in different mass loss of the latex. The temperature dependence for the rate of degradation can be expressed by the Arrhenius equation [6]:

$$k = Ae^{-E_a/(RT)} \tag{1}$$

where k = reaction rate constant, E_a = activation energy, A = collision frequency factor, R = gas constant and T = absolute temperature.

According to South et al. [18], as thermal ageing proceeded, the percentage of polysulphidic cross-links decreased and the percentage of monosulphidic cross-links increased which correlated with the higher mechanical property values. Another research by Barker et al. [19] concludes that temperature changes influence the rate of degradation and the alteration of the cross-linking system manages to maintain the material performance over a wide temperature range. Mott et al. [20] found that highly oxidised layer can develop on the surface of rubber and the degradation of thicker articles may be governed by diffusion instead of oxidation. Oxidation is the most important process for the material that contributes towards ageing.

The studies above were basically done to predict the effect of ageing on the material performances. The objective of this study is to investigate the effect of heat ageing under controlled and environmental condition on the puncture strength of UD-UHMWPE fabric coated with NRL.

Experimental Method

Materials

High strength UD-UHMWPE fabric with areal density of 150 g/m^2 was used in this study. For the coating material, high modulus pre-vulcanised NRL (supplied by Revertex Malaysia Sdn. Bhd.) with viscosity of about 25s (Ford cup 3 at 25 °C) and 60.5 % total solid content (TSC) was used.

Depth of Coating

Three coating depths or thicknesses were explored: single dip (SD), double dip (DD) and triple dip (TD). The UD fabric samples were impregnated into NRL and slowly taken out before hanging them to dry at room temperature. The second and third dipping were done after drying the samples for about 1 h, respectively.

Accelerated Heat Ageing

Controlled Condition

The purposes of the accelerated heat ageing treatment were to investigate the effect of heat on the puncture strength and to observe the changes that occur on the surface of the fabric. The heat ageing procedures took place in the laboratory using a cabinet oven with intermediate rate of fan flow according to the ASTMD3045-92 standard [21]. The UD samples were treated to temperatures of 40, 70 and 100 °C for 48 and 192 h. All of the UD fabrics were placed in a black polyester fabric to protect it from direct light and heat. After the treatment, the packs of uncoated and coated UD fabric were then conditioned in the laboratory for 24 h before the puncture strength test.

Environment Condition

The uncoated and coated UD-UHMWPE fabrics were placed inside a car for 8 months. The puncture strength test was conducted every 2 months interval starting at 2, 4, 6 and 8 months, respectively. The samples were covered with black polyester fabric to avoid direct exposure to light.

Puncture Strength

The puncture strength test was conducted using the testometric tensile tester following a modified puncture strength test as described by Ahmad et al. [3]. A single fabric sample was placed horizontally onto a circular clamp of 10 cm in diameter. The circular clamp has several fasteners (bolt) to grip the sample and prevent slippage during the test. A probe made of stainless steel having a dimension of 25.4 mm diameter and cone angle of 120° was used. During the test, the probe will move vertically downwards at a constant speed of 100 mm/min. The maximum force (N) to penetrate the sample indicates the puncture strength of the material [3].

FTIR Spectroscopy

The spectra of dried uncoated and coated UD-UHMWPE fabrics that had been aged at different elevated temperatures for 48 and 192 h were obtained using FTIR spectrophotometer with attenuated total reflectance (ATR) accessory. A total of 16 scans were accumulated with a resolution of 600 cm^{-1}.

Results and Discussion

Puncture Strength

The application of NRL on the UD-UHMWPE fabric results in the increment of the thickness of the material. The areal density of the materials increased as the thickness of the materials increased. In general, there is no significant difference between the thickness and areal density of the unaged and aged samples. Table 1 shows the physical properties of unaged sample. The puncture strengths for the SD-, DD- and TD-coated UD fabrics were 50, 66 and 87 % higher than uncoated UD fabric, respectively. This is due to the NRL and thickness of the materials which needed higher penetration force [2]. The flexibility of the NRL facilitates the force increment upon puncture by trapping the fibres within NRL layer. Unfortunately, the heat ageing process weakened the strength and the NRL loss its flexibility as shown in Fig. 1.

Figures 2 and 3 show the result of puncture strength at elevated temperatures for 48 and 192 h, respectively. It was observed that at 70 °C, the strength increased and then decreased significantly at 100 °C. These changes apply at both time intervals (48 and 192 h). This may be due to the cross-links of the materials. Based on prior investigations [4–12], it was determined that cross-linking improved the material's performance. However, further heating will increase crystallinity which results in the reduction of the impact resistance [6].

On the other hand, samples that were stored inside the car for the duration of 8 months did not show any significant difference. Figure 4 shows the temperature and relative humidity (RH) inside the car for the duration of 8 months according to Malaysia's tropical climate. The maximum temperature was 49 °C and minimum was 25 °C while maximum RH was 88 % and minimum was 48 %. In general, the average temperature was 32 °C and 48 % for RH. Table 2 shows the summary of the temperature and RH from 2 until 8 months.

From Fig. 5, it can be seen that the puncture strength generally increased for the first 2 months and decreased after that. This may be due to the effect of the heat and the humidity inside the car on the UD-UHMWPE fabric. It was observed that microorganisms grow on the surface of some samples which demonstrated the presence of moisture inside the car (Fig. 6). Cherian et al. [22] showed that Gram-positive bacteria were able to degrade natural rubber effectively. Another research [23] found that soil bacteria have high potential to degrade biodegradable NRL

Table 1 Physical properties of the unaged sample

Sample	Areal density (g/m^2)	Thickness (mm)
Uncoated	141	0.16
Single dip	230	0.26
Double dip	321	0.34
Triple dip	479	0.53

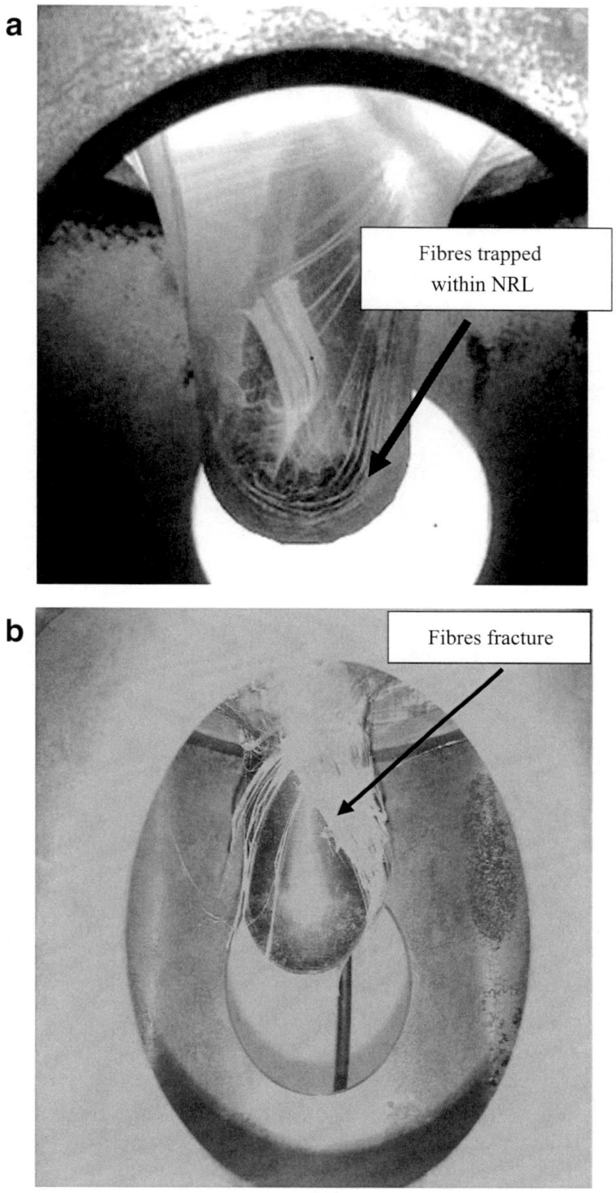

Fig. 1 Flexibility of the NRL before heat ageing (**a**) and after heated at 100 °C (**b**)

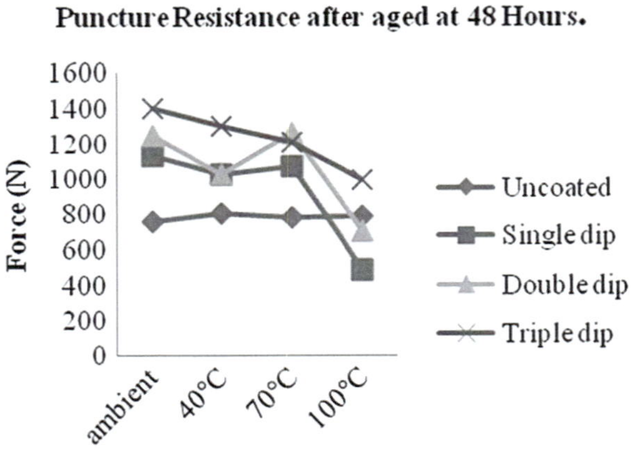

Fig. 2 Puncture strength after aged for 48 h at elevated temperature

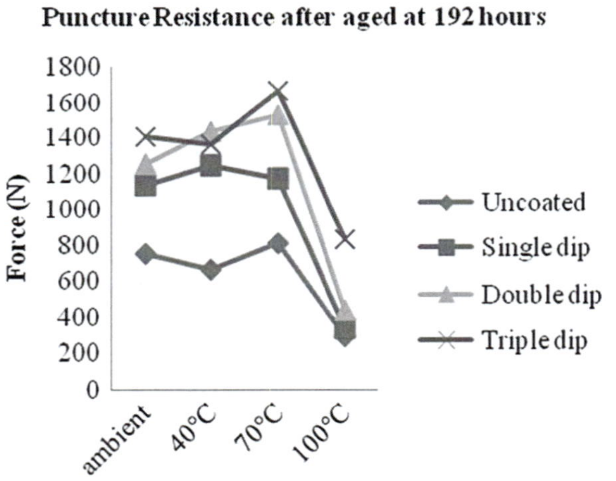

Fig. 3 Puncture strength after aged for 192 h at elevated temperature

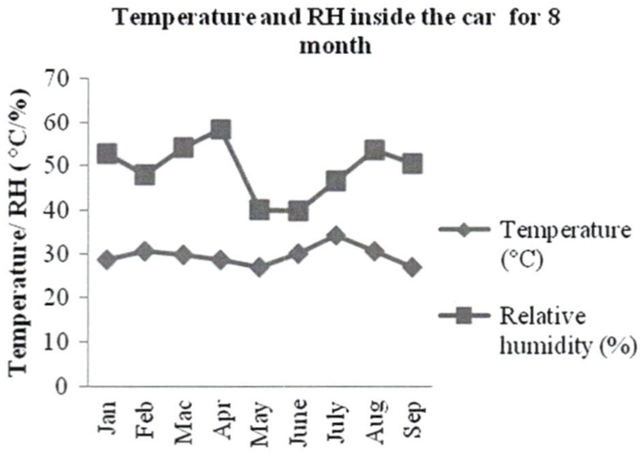

Fig. 4 Temperature and RH inside the car for 8 months

film by consuming the protein content as its sole carbon energy source. From previous investigations [13–16], humidity and microorganism have higher potential factor towards degradation of the NRL, thus weakening the strength of the materials.

Table 2 Temperature and RH inside the car

	Temperature (°C)		RH (%)	
Month	Max	Min	Max	Min
2	41	25	88	23
4	47	25	88	23
6	49	25	88	23
8	49	24	88	23

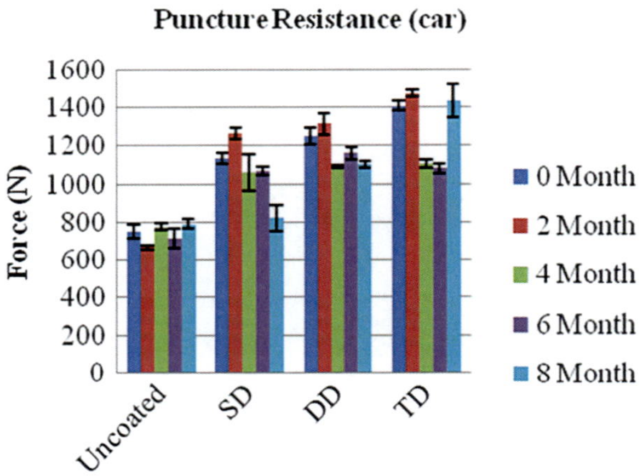

Fig. 5 Puncture strength of the samples inside the car for 8 months duration

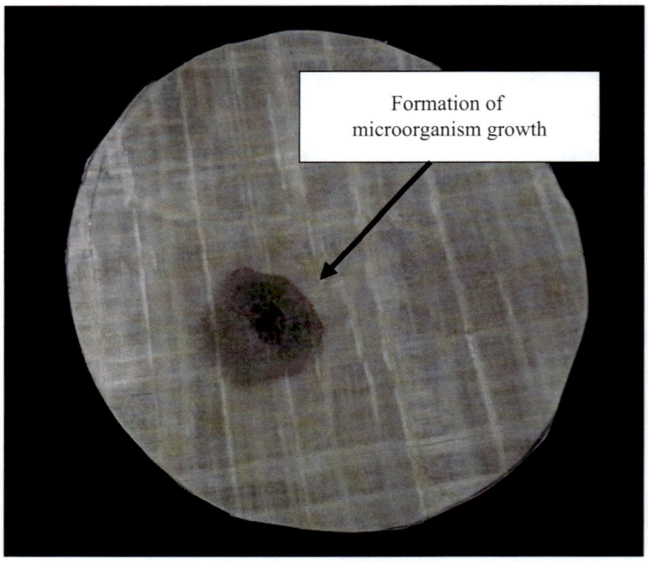

Fig. 6 Microorganism growth on the surface of NRL

FTIR Analysis

FTIR analysis was conducted in order to observe significant changes in the chemical structure of the material. The samples were divided under controlled and environmental condition for both uncoated and coated samples. It was observed that the pattern of the absorbance peaks of uncoated and coated samples (controlled condition)

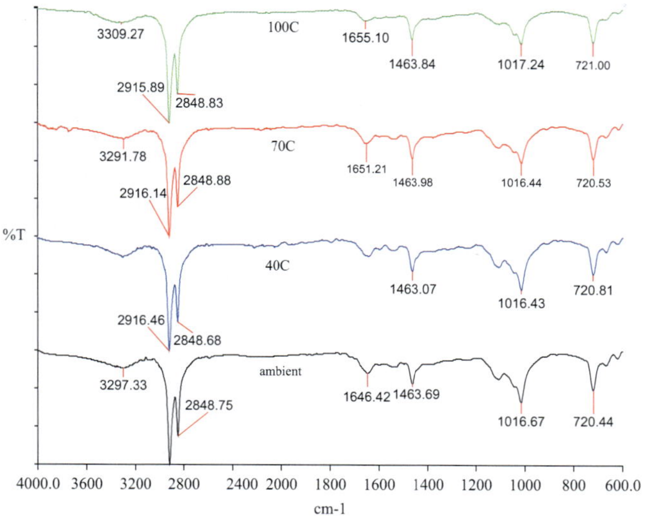

Fig. 7 FTIR spectra of uncoated samples at elevated temperatures (controlled condition)

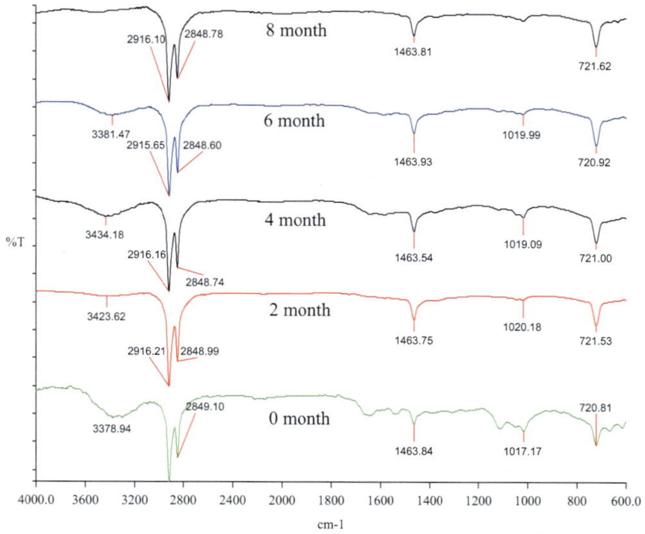

Fig. 9 FTIR spectra of uncoated samples for 8 months of heat storage (environmental condition)

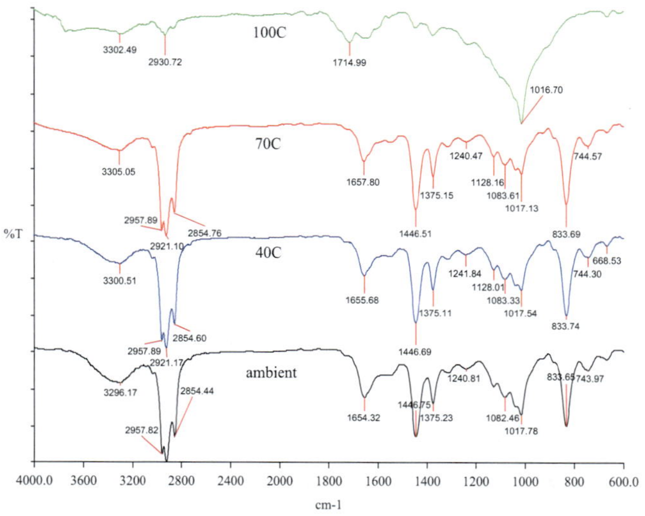

Fig. 8 FTIR spectra of coated samples at elevated temperatures (controlled condition)

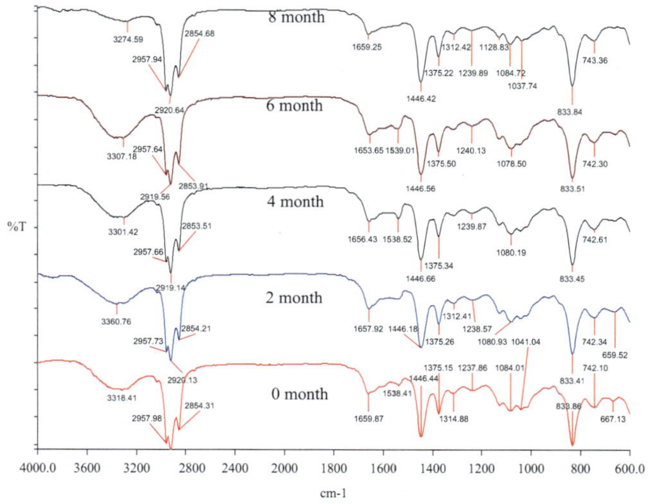

Fig. 10 FTIR spectra of coated samples for 8 months of heat storage (environmental condition)

appeared to be different because of the application of NRL on the surface of UD-UHMWPE fabric. From Fig. 7, it can be seen that there is no difference with the chemical structures at elevated temperature for uncoated samples. However, for the coated samples (Fig. 8), the peak from 1,083.61 cm^{-1} until 1,657.8 cm^{-1} disappeared and an absorbance peak at 1,714.99 cm^{-1} appeared after being exposed to temperature of 100 °C. These changes were observed for both the DD and the TD samples.

From the analysis, it can be said that the formation of the carbonyl groups, C=O, just started to appear. Oxidation takes place over a long period of time. Prolonged exposure with heat and humidity will lead to degradation where the decrease in C–O stretching mode will increase the C=O products with a further increase of ageing time [24].

In addition, FTIR spectra for samples stored inside the car (environmental condition) did not show any significant difference. The chemical structures remain the same for the 8-month storage. Figures 9 and 10 show the FTIR spectra of both uncoated and coated samples, respectively, for the 8-month storage.

Conclusion

The puncture strength increased with the application of NRL as the coating material on UD-UHMWPE fabrics in comparison with uncoated UD-UHMWPE fabric. Nevertheless, heat and humidity will reduce the performance of the material. It was observed that heat ageing at 70 °C increases the puncture strength of both uncoated and coated samples. However, samples that were exposed to environmental condition for a few months did not show any changes in terms of chemical structure but result in a reduction of the puncture strength. Further analysis will be conducted on DSC and TGA.

Acknowledgement The authors would like to thank the Research Management Institute (RMI) of Universiti Teknologi MARA for the research grant provided under the Research Cluster Fund (RCF).

References

1. M. Karahan, Comparison of ballistic performance and energy absorption capabilities of woven and unidirectional fabrics. Text. Res. J. **78**(8), 718–730 (2008)
2. N. Hassim, M.R. Ahmad, W.Y.W. Ahmad, A. Samsuri, M.H.M. Yahya, Puncture resistance of natural rubber latex unidirectional coated fabrics. J. Indus. Text. **42**(2), 118–131 (2012)
3. M.R. Ahmad, W.Y.W. Ahmad, A. Samsuri, J. Salleh, Ballistic response of natural rubber latex coated and uncoated fabric systems. J. Rubb. Res. **10**(4), 207–221 (2007)
4. D.D. Heckelman, Depth dependent oxidation and wear resistance of gamma irradiated UHMWPE as a function of aging condition. Ph.D. thesis, Rensselaer Polytechnic Institute, New York, 2011
5. S.J. Gencur, Fracture toughness and fatigue crack propagation behavior of conventional and highly crosslinked, thermally treated ultra high molecular weight polyethylene (UHMWPE). M.Sc. Thesis, Case Western Reserve University, 2004
6. W. Montague Jr., Cross-linked UHMWPE: process improvement-elimination of oxidation. M.Sc. Thesis, University of Massachusetts Lowell, 2008
7. F.E. Kennedy, B.H. Currier, J.H. Van Gitters, J.H. Currier, J.P. Collier, Oxidation of ultra high molecular weight polyethylene and its influence on contact fatigue and pitting of knee bearings. Tribol. Trans. **46**(1), 111–118 (2003)
8. V. Premnath, A. Bellare, E.W. Merrill, M. Jasty, W.H. Harris, Molecular rearrangements in ultra high molecular weight polyethylene after irradiation and long term storage in air. Polymer **40**, 2215–2229 (1999)
9. L. Costa, M.P. Luda, L. Trossarelli, Ultra high molecular weight polyethylene- I mechano- oxidative degradation. Polym. Degrad. Stab. **55**, 329–338 (1997)
10. L. Costa, M.P. Luda, L. Trossarelli, Ultra high molecular weight polyethylene- II thermal and photo oxidation. Polym. Degrad. Stab. **58**, 41–54 (1997)
11. P. Braco, V. Brunella, M.P. Luda, M. Zanetti, L. Costa, Radiation induce crosslinking of UHMWPE in the presence of co-agent: chemical and mechanical characterisation. Polymer **46**, 10648–10657 (2005)
12. M. Rocha, A. Mansur, H. Mansur, Characterisation and accelerated ageing of UHMWPE used in orthopedic prosthesis by peroxide. Materials **2**, 562–576 (2009)
13. K. Fortuniak, G. Pedlich, M. Landwijt, Effect of accelerated ageing conditions on the protective properties of anti blow product. Fibres Text. East. Eur. **19**(2), 61–64 (2011)
14. M. Fejdys, M. Landwijt, M.H. Struszczyk, Effect of accelerated ageing conditions on the degradation process of dyneema polyethylene composites. Fibres Text. East. Eur. **19**(1), 60–65 (2011)
15. A.A. Obaid, J.M. Deitzel, J.W. Gillespie Jr., J.Q. Zheng, The effect of environmental conditioning on tensile properties of high performance aramid fibers at near ambient temperatures. J. Compos. Mater. **45**(11), 1217–1231 (2011)
16. A.L.D.S. Alves, L.F.C. Nascimento, J.C.M. Suarez, Influence of weathering and gamma irradiation on the mechanical and ballistic behavior of UHMWPE composite armor. Polym. Test. **24**, 104–113 (2005)
17. D.L.S. Agostini, C.J.L. Constantino, A.E. Job, Thermal degradation of both latex and latex cast films forming membranes: combined TG/FTIR investigation. J. Therm. Anal. Calorim. **91**, 703–707 (2008)
18. J.T. South, S.W. Case, Thermal ageing on the quasi-static properties of NR. Natuurrubber **36**(4) (2004)
19. L.R. Barker, Accelerated and long term ageing of NR vulcanizates. Natuurrubber **36**(4) (2004)
20. P.H. Mott, C.M. Roland, Effect of air and seawater oxidation on the mechanical failure of NR. Natuurrubber **36**(4) (2004)
21. ASTM D3045-92, *Standard Practice for Heat Aging of Plastics Without Load* (American Society for Testing Materials, West Conshohocken, PA, 2010)
22. E. Cherian, K. Jayachandran, Microbial degradation of natural rubber latex by a novel species of Bacillus sp. SBS^{25} isolated from soil. Mater. J. Environ. Res. **3**(4), 599–604 (2009)
23. Z.M.F. Tajuddin, A.R. Azura, *Bacillus megaterium* sp. as degradation agent for biodegradable natural rubber latex films. Adv. Mater. Res. **626**, 813–817 (2013)
24. D. Dazylah, M.S. Ma'zam, M.Y. Amir Hashim, Change in the micro structure and tensile properties of peroxide prevulcanised natural rubber latex films during thermal oxidative ageing. J. Rubb. Res. **10**(4), 222–254 (2007)

Modeling Plain Woven Composite Model with Isotropic Behavior

M.F. Yahya, S.A. Ghani, and J. Salleh

Abstract

Research in woven fabric composites modeling and simulation has been extensively done in the past decade. The simulation issue is associated with estimating yarn transverse isotropic property. Most simulation models are attempting to achieve high level of accuracy with unit cell approach. Thus, in the present study, an alternative finite element analysis approach was designed to model uniaxial tensile performance of Kevlar-based plain woven fabric with isotropic material property. The dynamics of a finite element model were investigated at a wide range of mesh levels. The results show that the plain woven fabric failure critically depended on the yarn and woven fabric structure. Good comparisons were achieved with experimental work, and further application in more complex weaves is suggested.

Keywords

Finite element analysis • Isotropic • Plain woven fabric • Uniaxial tensile • Composites

Introduction

Over the years, considerable efforts were done to use textile structures for technical textile application, particularly in composites. Strength-to-weight and stiffness-to-weight ratios are some of the benefits offered by textile materials as composite precursor [1, 2]. Textile composites consist of two distinct components that are mixed by composite fabrication process [3]. Those components are textile reinforcement and matrix. Specific performance of textile composites can be achieved by tailoring textile properties and specific cost in mind [4]. The mechanical properties of textile structures and its composites must be fully understood before submitting them for real-life applications. Simulation analysis technique offers an opportunity to predict the behavior of textile structures before the actual product can be produced. Finite element analysis is one of the simulation packages that are available today. Finite element analysis requires textile structures to be visualized and analyzed in the form of geometric representation, either in the form of 2D, 3D, shell, or even pin jointed.

Literatures on plain woven composites are extensive [5–12]. Most of the articles focused upon glass- and carbon-based woven fabric for biaxial tensile. To the best of authors' knowledge, Duan [13–15] reported extensive simulation work on tensile and impact behavior of Kevlar-based woven fabric. Other experimental works are from Luo [16], Sun [17], and Wang [18]. However, according to these papers, impact is the most investigated mechanical properties of transverse isotropic Kevlar dry fabrics and composites. Although their models predict excellent results when compared with practical works, still, the numerical approach for determining transverse isotropic is inconsistent to many researchers. The deviation is due to issues in determining the transverse modulus [19] and numerical convergences [20].

The highlight was on plain woven fabric composite model development based on physical fabric properties.

The research was funded with Fundamental Research Grant Scheme (FRGS) 2011–2013 from the Ministry of Education, Malaysia.

M.F. Yahya (✉) • S.A. Ghani • J. Salleh
Faculty of Applied Sciences, Universiti Teknologi MARA (UiTM), 40450 Shah Alam, Selangor Darul Ehsan, Malaysia
e-mail: faizulyahya@salam.uitm.edu.my

M.R. Ahmad and M.F. Yahya (eds.), *Proceedings of the International Colloquium in Textile Engineering, Fashion, Apparel and Design 2014 (ICTEFAD 2014)*, DOI 10.1007/978-981-287-011-7_4,

Other modeling issues accompanying 3D mechanical model were also highlighted. For this reason, the objectives of the study were to review the effect of the isotropic material assumption in plain woven fabric models.

Finite Element Modeling

Woven Fabric and Resin Geometrical Model

A 3D woven fabric model was generated with Abaqus preprocessor CAD tool. Woven fabrics composed of yarns arranged 90° to each other and often referred as warp and weft yarns. Yarn geometry model formation is based on yarn cross-sectional shape and yarn path. The yarn cross-sectional shape was evaluated as lenticular with digital stereo microscope with width A at 1.43 mm and height B at 0.4 mm. These measurements yield four coordinate points for describing yarn cross-sectional shape. These coordinates were adopted to formulate yarn lenticular cross-sectional shape equation. Yarn path was formed by considering yarn width A and yarn height B effects. Lenticular-shape formulation to describe yarn cross section and paths is presented in Eqs. (1) and (2). A woven geometrical model was constructed with four warp and weft yarns. The yarns were in lenticular cross-sectional shape and assumed in monofilament form. The procedures for the yarn geometrical model development were described in earlier publications [21, 22]. The matrix was designed with Abaqus preprocessor CAD program. Matrix dimensions were determined by yarn model volume V_y and fiber volume fraction V_f parameters. Matrix volume, V_m in mm^2, was obtained according to Eq. (3). The matrix dimensions are summarized in Table 1.

$$-0.41x^2 + 0.2 = y \tag{1}$$

$$-2.48x^2 + 0.2 = y \tag{2}$$

Warp, weft yarns, and matrix geometrical models were assembled to form the fourth model component, which is the cut matrix model. This procedure was critical for efficient geometry contact and interaction in the analysis. Yarns and cut matrix geometrical models are presented in Figs. 1 and 2.

$$V_m = 100 \cdot \frac{V_y}{V_f} - V_y \tag{3}$$

Material Properties

The yarn material property was determined experimentally using uniaxial tensile test. The test was performed based on BS EN ISO 2062:1995 standard. The specimen length required for the test was 250 mm at 100 mm m^{-1} crosshead speed. The test yields yarn modulus at 19 GPa. Similar to previous work on dry fabric simulation, warp and weft yarn properties in composite structure were adopted as isotropic, especially in the elastic region. The deformation in this region is completely recoverable if the load is removed. In this region, material stiffness behavior or modulus can be determined by calculating the ratio between stress and strains. The yarn Poisson's ratio was determined separately with yarn experimental and simulation validation result.

Table 1 Matrix dimensions

Parameters	Units
Yarn volume, V_y mm^3	6.07
Fiber volume fraction, V_f	37.2
Matrix volume, V_m mm^3	10.247
Matrix width, mm	7.211
Matrix length, mm	7.211
Matrix thickness, mm	0.8

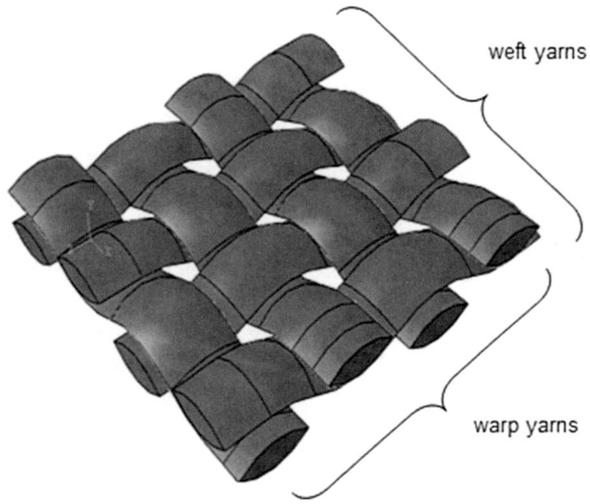

Fig. 1 Woven fabric model

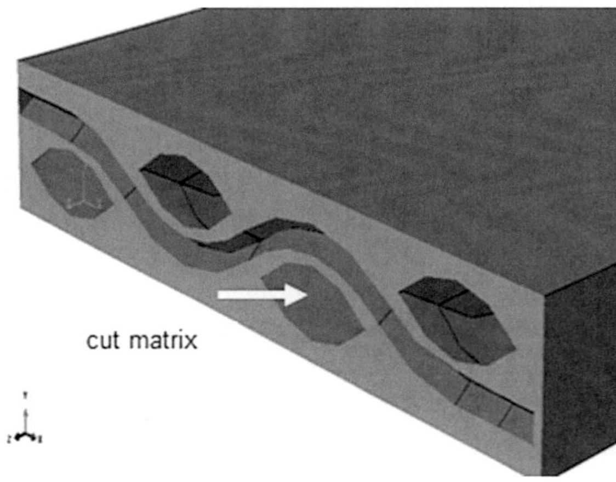

Fig. 2 Cut matrix model

Table 2 Material properties

	Modulus, E	Poisson ratio	Density, g cm^{-2}
Kevlar	19,000	0.37	1.44
Resin	3,450	0.35	1.093

Table 3 Yarn plastic properties

True stress (Pa)	Plastic strain (mm)
5.40E + 08	0
9.57E + 08	1.05800632
1.42E + 09	1.186822575
1.88E + 09	1.294996516

Yarn density was determined with BS EN ISO 1183-1:2004 standard. Resin parameters were chosen from manufacturer's technical data sheet. Yarn and resin material properties are presented in Table 2.

An additional procedure was required to convert yarn elastic to plastic behavior. The approach was necessary to consider yarn material property transition from elastic to plastic region, as it exceeded its yield point. Yarn experimental stress–strain was decomposed into true stress and plastic strains. The conversion process was performed with Eqs. (4) and (5). The resin was considered to have brittle property and failed once yield point surpassed. Yarn true stress and plastic strains are shown in Table 3.

$$\varepsilon^{\mathrm{pl}} = \varepsilon^{t} - \varepsilon^{\mathrm{el}} = \varepsilon^{t} - \frac{\sigma}{E} \tag{4}$$

$$\sigma = \sigma_{\mathrm{norm}}(1 + \varepsilon_{\mathrm{norm}}) \tag{5}$$

where $\varepsilon^{\mathrm{pl}}$ is true plastic strain, ε^{t} is true total strain, $\varepsilon^{\mathrm{el}}$ is true elastic strain, σ is true stress, E is modulus, σ_{norm} is nominal stress, and $\varepsilon_{\mathrm{norm}}$ is nominal strain.

Interactions, Loading, and Boundary Conditions

Both yarns and resins were fused together during simulation analysis. The type of contacts replicated yarn and resin interactions of composite once curing process is completed. Due to loading direction, the warp yarn model was assigned to behave as master, while weft and resins were assigned to slave contact surface. The normal stress acting on the composites was recalculated by considering the matrix area in mm^2 as in Eq. (6). Table 4 summarized the stress required for uniaxial tensile simulation.

$$\text{Stress (MPa)} = \frac{\text{Load (N)}}{\text{Area}\,(\text{mm}^2)} \tag{6}$$

Boundary condition was applied in finite element simulation for restraining unloaded composite end in longitudinal and rotational directions. The approach attempted to replicate actual constraint existed during uniaxial tensile test. The composite model was restrained to move in all principal directions (x, y, z) as well as torsion forces (UR1, UR2, UR3). With the approach, top section of the composite model was fixed in its position during the analysis. Similar to dry fabric model, warp yarn was assigned to only move in one direction or U2. Other directions are kept zero or constraint at all computation time. The procedures are illustrated in Fig. 3a, b.

Table 4 Matrix area

Composite thickness (mm)	0.62
Composite width (mm)	25
Load (N)	50,000
Stress (MPa)	3,225.81

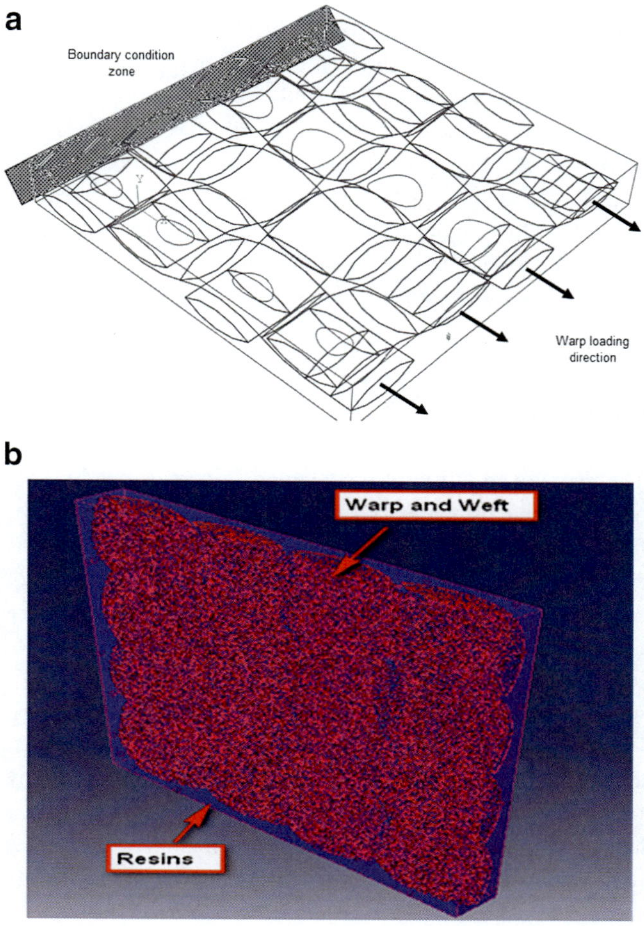

Fig. 3 (**a**) Boundary condition and loading (**b**) yarn interactions

Mesh and Mesh Convergence

Warp and weft yarns were meshed into four-node tetrahedron shape. A total of 12° of freedom was assigned to all the nodes. The tetrahedron element shape was selected because of the yarn nonlinearity geometrical shape. The matrix was meshed with hexahedron element shape. Mesh is a process to break larger 3D geometrical model into smaller elements for analysis. At each element, local displacement, strains, and stress are initially calculated before global stress-strain is determined. Mesh convergence was conducted to evaluate the effect of mesh fineness towards stress–strain results. In the test, total elements in simulation models systematically increased from coarser to finer elements. Four mesh levels were selected for the mesh convergence analysis. The levels were 7,500, 15,000, 30,000, and 60,000 elements. These mesh sizes were set up to review the model validation accuracy with respect to experimental results. The mesh convergence and model validation results were discussed in the following section.

Composite Fabrication

The dry woven fabric was converted to composites by vacuum bagging principle. Vacuum infusion process was preferred when a large structure with low throughput time is considered. The system offers a reduction of top presser tool cost and the ability to control styrene vaporization. Sample thickness and fiber volume fractions were controlled effectively. Araldite resin LY 5052-1 and Aradur hardener 5052-1 were mixed according to 62:38 resin to hardener ratios. Woven fabrics were carefully laid up inside the sealed vacuum bag. Resin moved steadily into the bagging system by vacuuming the air out. Efficient seal was needed to prevent leakage to vacuum bagging unit. Woven fabric composites were removed from vacuum bagging unit after 24 h to allow sufficient gelation time.

Tensile Experiment

The composite tensile test provided information about its resistance to fail during uniaxial tension. The uniaxial tensile load was applied in warp direction. EN ISO 527-4:1997 standard was used to test the composite samples. The standard load for the test was 50 kN and performed at 2 mm min^{1} crosshead speed. Specimen dimension has 250 mm length and 25 mm width. Sample slippages were solved by assembling additional four steel tabs on all sample edges front and back faces. Steel tabs were cut according to specimen dimension. Strong steel clamps and sample bonding were achieved by using identical resin and hardener as in composite fabrication.

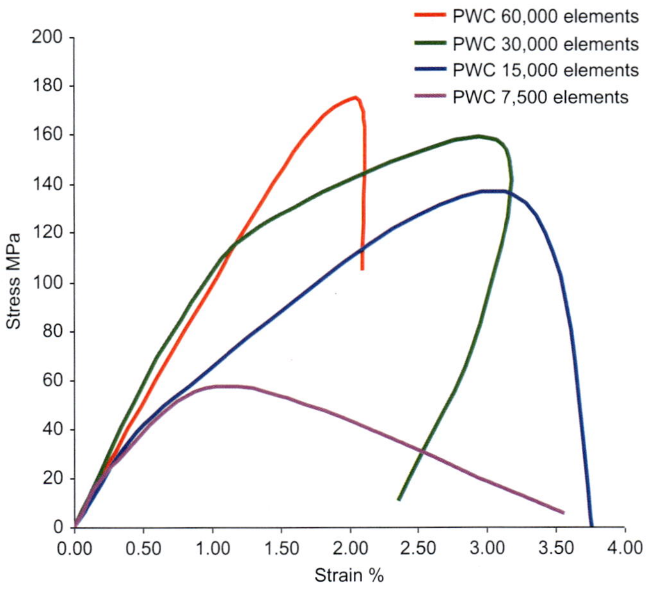

Fig. 4 Mesh convergence result

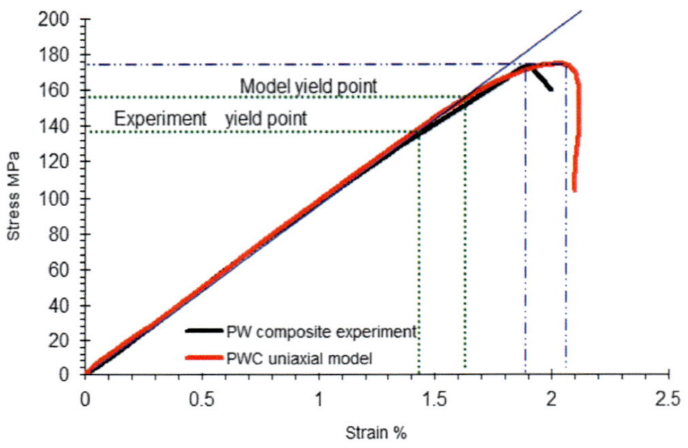

Fig. 5 Simulation and experimental validation

Results

Mesh Convergence

Coarser mesh size was represented by smaller element numbers existed in the model and vice versa. It was shown that in finest element mesh sizes, 60,000 elements had presented the highest stress result as compared to other mesh categories. Finest meshes at 60,000 total elements presented closest results to practical and therefore selected for analysis. Mesh convergence test results are outlined in Fig. 4. The increase of meshes had shown to improve model response in the elastic region. Nevertheless, the finer mesh sizes had also resulted poor computation and analysis time.

Table 5 Uniaxial strength validation

Elastic modulus	GPa	Stress yield point	MPa	Strain yield point	%
Simulation (GPa)	9.619	Simulation (MPa)	156	Simulation (%)	1.61
Experimental (GPa)	9.702	Experimental (MPa)	136	Experimental (%)	1.42
Simulation error (%)	−0.855	Simulation error (%)	14.7	Simulation error (%)	13.4

Simulation and Experimental Validation

Simulation and experimental results of uniaxial tension performance of plain woven fabric composite are presented in Fig. 5. Satisfactory elastic modulus validation agreement was achieved with both simulation and experimental stress–strain results. The error difference between those two curves is presented in Table 5. Simulation error was considered by simulation and experimental modulus differences in percentages. Plain woven composite showed modulus error at −0.855 %. Both stress and strain simulation errors were represented by 14.7 and 13.4 %, respectively. The result suggested that composite performance during initial loading stage was well represented by yarn and resin elastic modulus. However, the simulation failure behavior was higher than the experimental results. This was because the material behavior was set up as isotropic. Yarn and matrix produced higher results in all directions. Material behavior in transverse, E_2, and compression, E_3, is usually lower in comparison to longitudinal modulus, E_1. Differential modulus in three primary material directions and shear effects during failure were not properly expressed with isotropic material behavior.

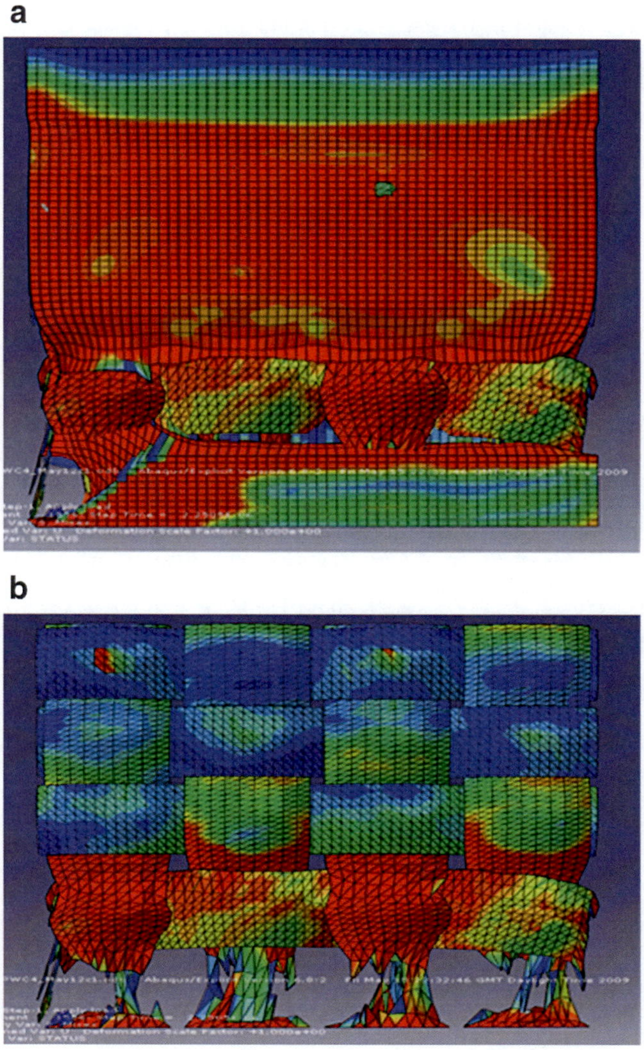

Fig. 6 Damage evolution (**a**) composite (**b**) fabric only

Damage Evolution

The damage evolution of plain woven composite is illustrated in Fig. 6a, b. It was shown that composite failure behavior was influenced by yarn damage. Necking and yielding effects were observed as composite damage initiated. Matrix deformations started when the damaged elements were removed from the finite element computation. At this stage, matrix elements were considered failed and matrix started to transfer its load to yarns. This effect was illustrated as matrix model extracted from the composite model as shown in Fig. 6b. At this stage, 90 % matrix elements failed and warp yarns responded to the uniaxial tension. Here, warp yarns were stretched further and finally failed as it surpassed breaking point. It was clear that composite damage mechanism was influenced by higher modulus yarn strength rather than matrix.

Conclusion

A simulation analysis procedure was developed to model plain woven fabric composites based on isotropic material property. Geometrical models were based on lenticular yarn cross-sectional shape. Elastic modulus of woven fabric was obtained using finite element analysis. The values predicted by the system compared favorably with experimental results, especially in elastic phase. The model is not complex, as it is based on physical fabric representation. The developed finite element procedure suggested further composite modeling with complicated weave structures such as twill, satin, and multilayer.

Acknowledgment Support for this research work provided by Ministry of Education through the Fundamental Research Grant Scheme (FRGS).

The authors acknowledged the assistance from the Research Management Institute (RMI) of Universiti Teknologi MARA, Malaysia.

References

1. H. Gu, Tensile and bending behaviours of laminates with various fabric orientation. Mater. Des. 1–4 (2005)
2. I. Ivanov, A. Tabiei, Three-dimensional computational micro-mechanical model for woven fabric composites. Compos. Struct. **54**, 489–496 (2001)
3. F.L. Matthews, G.A.O. Davies, D. Hitchings, C. Soutis, *Finite Element Modelling of Composite Materials and Structures*, 1st edn. (Woodhead Publishing Limited, Cambridge, 2000)
4. F.K. Ko, in *3-D Textile Reinforcements in Composite Materials*, ed. by A. Miravat, vol. 1 (Woodhead Publishing Limited, Boca Raton, 1999)
5. B.N. Cox, M.S. Dadkhah, W.L. Morris, On the tensile failure of 3D woven composites. Compos. Part A **27A**, 447–458 (1995)
6. D. Goyal, J.D. Whitcomb, Load flow in plain woven composites. J. Compos. Mater. **42**, 2761–71 (2008)
7. G. Hivet, P. Boise, Consistent 3D geometrical model of fabric elementary cell. Application to a meshing preprocessor for 3D finite element analysis. Finite Element Anal. Des. **42**, 25–49 (2005)
8. W. Warren, The large deformation elastic response of woven kevlar fabric. Polym. Compos. **13**, 278–284 (1992)
9. J. Whitcomb, X. Tang, Effective moduli of woven composites. J. Compos. Mater. **35**, 2127–2144 (2001)
10. G. Hivet, P. Boisse, Consistent geometrical description and finite element modeling of woven fabrics. Application to compaction. Finite Element Model. Text. Text. Compos. 1–4 (2007)
11. E.H. Glaessgen, C.M. Pastore, O.H. Griffin, A. Birger, Geometrical and finite element modelling of textile composites. Compos. Part B **27B**, 43–50 (1996)
12. I.M. Daniel, J.J. Luo, P.M. Schubel, Three-dimensional characterization of textile composites. Compos. Part B **39**, 13–19 (2008)
13. Y. Duan, M. Keefe, T.A. Bogetti, B. Powers, Finite element modeling of transverse impact on ballistic fabric. Int. J. Mech. Sci. **48**, 33–43 (2006)
14. Y. Duan, M. Keefe, T.A. Bogetti, B.A. Cheeseman, B. Powers, A numerical investigation of the influence of friction on energy absorption by a high-strength fabric subjected to ballistic impact. Int. J. Impact Eng. **32**, 1299–1312 (2006)
15. Y. Duan, M. Keefe, T.A. Bogetti, B.A. Cheeseman, Modeling the role of friction during ballistic impact of high-strength plain-weave fabric. Compos. Struct. **68**, 331–337 (2005)
16. Y. Luo, L.V. Lihua, B. Sun, Y. Qiu, B. Gu, Transverse impact behaviour and energy absorption of three-dimensional orthogonal hybrid woven composites. Compos. Struct. **81**, 202–209 (2006)
17. B. Sun, Y. Li, B. Gu, A unit cell approach of finite element calculation of ballistic impact damage of 3-D orthogonal woven composite. Compos. Part B **40**, 552–560 (2009)
18. X. Wang, B. Hu, Y. Feng, F. Liang, J. Mo, J. Xiong, Y. Qiu, Low velocity impact properties of 3D woven basalt/aramid hybrid composites. Compos. Sci. Technol. **68**, 1–7 (2007)
19. M. Valizadeh, S. Lomov, S.A.H. Ravandi, M. Salimi, S.Z. Rad, Finite element simulation of Yarn Pullout test for plain woven fabrics. Text. Res. J. **80**, 1–12 (2009)
20. A. Gasser, P. Boisse, S. Hanklar, Mechanical behavior of dry fabric reinforcements. 3D simulations versus biaxial test. Comput. Mater. Sci. **17**, 7–20 (2000)
21. M.F. Yahya, J. Salleh, W. Wan Ahmad, Uniaxial failure resistance of square-isotropic 3D woven fabric modelled with finite element analysis. Business, Engineering and Industrial Applications (ISBEIA), 2011 IEEE Symposium on: IEEE (2011), pp. 16–21.
22. M.F. Yahya, S.A. Ghani, J. Salleh, Effect of impactor shapes and yarn frictional effects on plain woven fabric puncture simulation. Textile Res J. **84**, 1095–105 (2014)

The Effect of Fabric Weave on the Tensile Strength of Woven Kenaf Reinforced Unsaturated Polyester Composite

M.P. Saiman, M.S. Wahab, and M.U. Wahit

Abstract

Natural fibers have gained attention in composite making since they are sustainable, renewable, and environmental friendly. However, there are some drawbacks such as low tensile strength, modulus, and flexural strength compared to industrial synthetic fibers. Various techniques have been used to increase the mechanical properties of the reinforced material to compete with synthetic fibers. This study produces reinforced fabric made from kenaf fibers woven into four different weave patterns. The reason for the use of different weave patterns is to reduce the crimp percentage as it may increase the tensile strength of the fabric. The weave patterns are Twill 4/4, Satin 8/3 and Basket 4/4. Plain 1/1 is used as the benchmark. The dry fabric structure with different weave patterns was optimized using a simulation of WiseTex software. The dry fabrics were infused with unsaturated polyester (UPE) to produce composite panel using vacuum infusion process (VIP). The dry fabric and the composites were tested for tensile strength and compared with the plain 1/1 weave pattern. The results show that the breaking strength of dry fabrics increased when different weave patterns with low crimp percentages were used compared with Plain 1/1. However, in composite form, there is a reduction in the tensile strength of Basket 4/4. The other weave patterns still retain the trend of increment in tensile strength compared with Plain 1/1 weave.

Keywords

Fabric design • Textile composite • Kenaf • Tensile strength • Vacuum infusion process

Introduction

In recent years, interest in woven fabric made from synthetic fibers for textile composite has risen. This is due to ease of handling, low cost, balanced in-plane properties, good impact resistance, and damage tolerance. Nowadays, researchers are now attracted to use natural fibers such as Kenaf, ramie, and sisal as reinforced materials in textile composites. Although synthetic fibers meet structural and durability demands, the advantages of natural fibers are as follows: low cost, low density, comparable specific strength, ease of separation, and can be recycled [1]. Study to replace synthetic fibers in composite due to its environmental issues and potential health hazard has resulted in the increased demand for natural fibers.

Designing a dry fabric for textile composite is greatly influenced by several parameters such as the fibers, yarn size, fabric count and the weave design. Strength of fabrics is due to the conversion of fiber strength to yarn strength and finally to fabric strength. When a load is applied on a woven fabric, frictional resistance between fibers occurs, followed by crimp interchange, and lastly the yarn extension where the load is applied towards the yarns [2]. Yarn crimp is the

M.P. Saiman (✉) • M.S. Wahab
Department of Manufacturing and Industry, Universiti Tun Hussein Onn Malaysia, Johor, Malaysia
e-mail: pahmi.saiman@gmail.com

M.U. Wahit
Center for Composites, Universiti Teknologi Malaysia, Johor, Malaysia

M.R. Ahmad and M.F. Yahya (eds.), *Proceedings of the International Colloquium in Textile Engineering, Fashion, Apparel and Design 2014 (ICTEFAD 2014)*, DOI 10.1007/978-981-287-011-7_5,

waviness of warp yarn and weft yarn interlacing together to produce fabric construction. It is affected by yarn count, fabric structure, and weaving tensions and related to the strength of textile fabric. If a load is applied on a woven fabric and the yarns are straight and non-crimp, the full load will be faced in tension at complete strength. However, if the yarns are crimped or bent, then the initial load will be consumed in straightening bent tows and then take up load subsequently leading to low strength materials [3].

There are several fabric weaves being used in textile composites such as glass, carbon, and aramid fibers. The fabric weave is determined by the technique of the interlacement between warp and weft yarns. The Plain weave is the basic reinforcing structure in textile composite, stable but has high crimp percentage. The Basket weave is similar to the plain weave structure, except in the alternate interlace yarns are two or more warp and weft yarns. It is much stronger and more pliable than Plain weave but less stable. The Twill weave has a much looser structure and is characterized by a diagonal line created by floated yarns. It is more pliable than the Plain weave, offers better drapability, and is more stable compared to Satin weave. Satin weave has a smooth surface with lots of floated warp yarns, good drapability, very pliable weave, and used for forming curved surfaces [4].

Study has shown that during the weaving process, fabric contraction occurs at the fabric fell causing changes in the fabric structure. It is also related with the fabric weave and the changing of the structure which has significant influence on some fabric properties such as the porosity, air permeability, thickness, crimp, elongation, and strength [5]. The changing of the fabric structure may cause a jammed structure. In this situation, warp and weft yarns do not have mobility within the structure as they are in intimate contact with each other [6]. All the factors can affect the properties of dry fabrics especially after converting them into a natural-based composite.

Unsaturated polyester (UPE) is a matrix resin usually used as a binder in fiber reinforced polymer (FRP) products. It is one of the most popular thermosetting resins due to its low cost, excellent process ability, and good cross-linking tendency with good mechanical properties when cured [7]. Due to the properties of cure capability at room temperature and low viscosity, it can combine with natural fibers and is widely used in industrial applications compared to other thermosetting resin [8]. The process ability at low pressures has made it easy to handle in processes like hand layup, filament winding, and liquid molding composite.

The objective of this paper is to investigate and analyze the influence of weave design towards woven kenaf dry fabric as structural material for the fabrication of natural-based composite. The investigation is on the yarn crimp which is related with the tensile strength of the dry fabric and in composite form. The stability of the fabric structure for different weave designs is also a concern as it may affect the tensile property of the composite.

Experimental

Materials

The Kenaf yarn was supplied by Juteko Bangladesh Pvt. Ltd., Dhaka, Bangladesh, with linear density of 759tex. UPE resin (2597P-I) was used as the matrix material, and methyl ethyl ketone peroxide (MEKP) as the catalyst was provided by Wee Tee Tong Chemical Pvt. Ltd., Singapore.

Preparation of Woven Kenaf

The Kenaf yarn was woven into four different weave structures: Plain 1/1, Twill 4/4, 8 harness Satin and Basket 4/4 using a floor loom (Fig. 1). The weave structures were selected as they are commonly used for industrial purposes [9–11]. Each selected weave has a long float which will reduce the yarn crimp percentage, and this will give a clear effect when compared to Plain weave. The design is also compatible for processing using the floor loom.

The fabric count was predicted using WiseTex software to produce a non-jammed balanced dry fabric suitable to be processed into composite. The result after weaving process shows that the plain fabrics are unbalanced fabric as the weft direction is much lesser than the warp direction. The other weave designs are successfully being made into balance fabric.

Fabrication of Composite Panels

The four different weave designs were fabricated into composite panels using UPE as the matrix and processed with VIP. VIP is a closed mold with one side rigid and the other flexible which is usually covered by vacuum bagging.

Testing

Crimp Percentage

The crimp percentage is calculated by dividing the difference between straightened thread length and the distance between the ends of the thread in the fabric:

$$C, \% = (L - S) \div S \times 100 \tag{1}$$

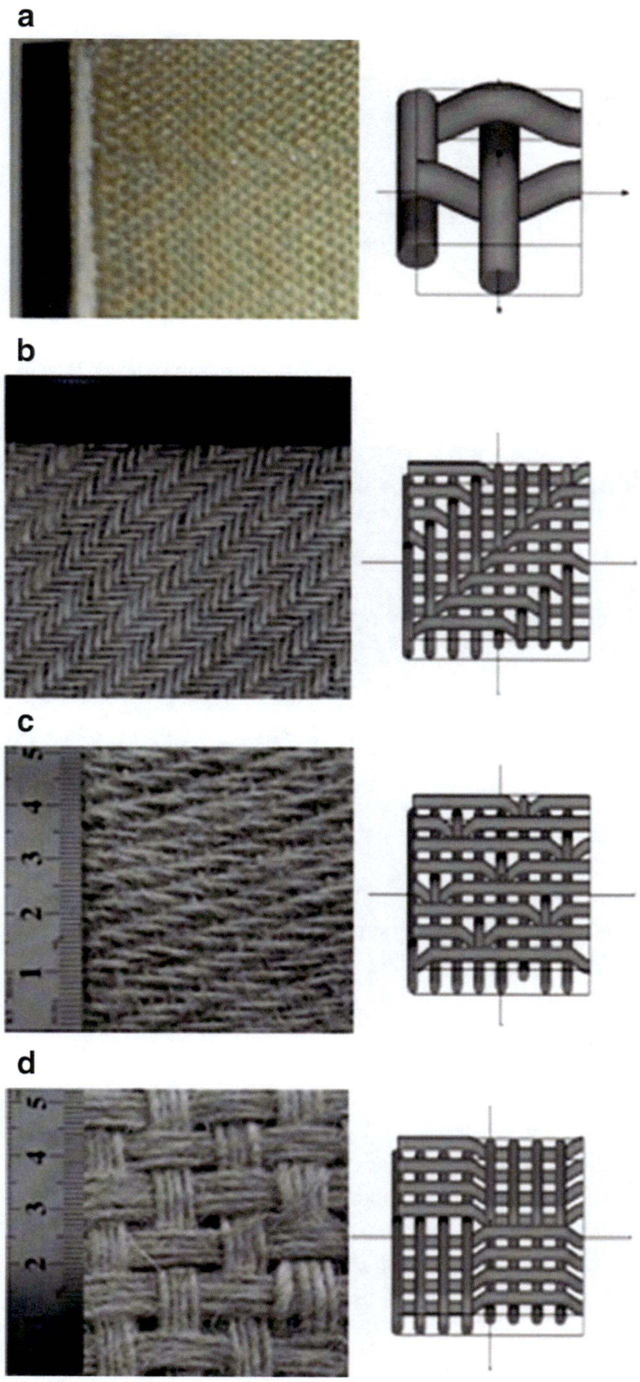

Fig. 1 Photographs and 3D images from WiseTex software of dry fabric: (**a**) Plain 1/1 (**b**) Twill 4/4 (**c**) Satin 8/3 (**d**) Basket 4/4

where C is crimp percentage, L is straightened yarn length, and S is fabric length.

Microscopy Tests

A Meiji Techno optic microscope equipped with a video camera and VIS PLUS version 2.0 software was used to take a digital image of the composite and the fracture surface.

Tensile Test

The tensile test was conducted using LLYOD, LR30K Universal Tensile Machine, and the test sample proceeds only in the warp direction. A revealed strip test was used to determine the breaking strength of the dry fabrics according to the standard test of ASTM 5035-95C. The loading rate for five specimens tested is at 300 $\pm$ 10 mm per min. The tensile test for the composite was carried out according to ASTM D3039, and each test used five specimens at a crosshead speed of 1 mm/min.

Result and Discussion

Effect of Crimp on a Dry Fabric

Table 1 shows four different weave designs with balanced fabric, except plain weave due to floor loom limitation. The yarn crimp at weft direction is much higher compared to warp direction for Twill 4/4, Satin 8/3 and Basket 4/4. It is different with Plain weave, where the yarn crimp in warp direction is much higher compared to weft direction. The difference in yarn crimp percentage in warp and weft direction may be due to high tension in one direction causing crimp interchange in the other direction [12]. Overall, the results show that Plain weave has higher crimp percentage compared to others. Theoretically, Plain weave will have high crimp due to its maximum interlacement compared to other weaves. Less crimp will result in greater strength on the fabric as it is free from inter-yarn and fibers contribute towards the force.

Figure 2 shows the breaking strength of dry fabric in warp direction for different weave designs. There are significant differences in crimp percentage and breaking strength of the fabric weave. The highest crimp percentage which is plain fabric has the lowest breaking strength. There is little difference in crimp percentage between Twill 4/4, Satin 8/3 and Basket 4/4, but the result of the breaking strength shows Basket 4/4 has the highest breaking strength followed by Satin 8/3 and lastly Twill 4/4.

Effect of Designs on a Composite

Figure 3 shows the tensile strength of the composite for the selected weave designs. The trend is similar with the breaking strength of the dry fabrics except basket 4/4 composite

Table 1 Fabric count and crimp percentage of selected weave design

Fabric specifications		Plain 1/1	Twill 4/4	Satin 8/3	Basket 4/4
Fabric count	Ends per cm	5	5	5	5
	Picks per cm	4	5	5	5
Crimp, %	Warp	16	3.0	3.7	2.8
	Weft	8.1	6.2	4.4	4.7

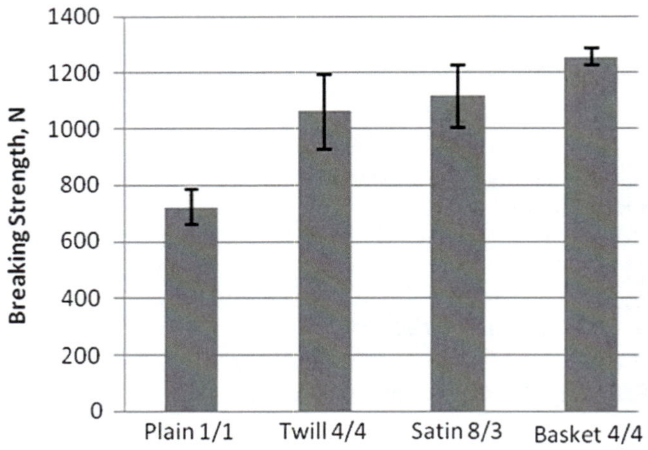

Fig. 2 Breaking strength of dry fabric in warp direction with different weave designs

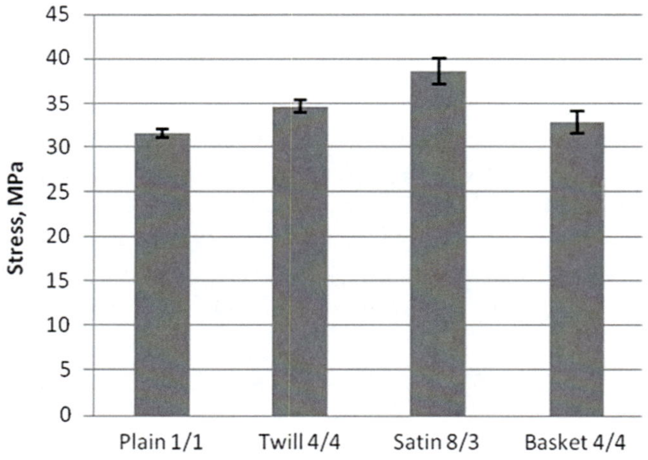

Fig. 3 Tensile strength of composite for different weave design

which shows a reduction in tensile strength when in composite form.

During the weaving process of the dry fabric, contraction occurs on the cloth fell or the reed of the loom [5]. This causes the floating adjacent yarns which are free from interlacement to join together to form jammed structure

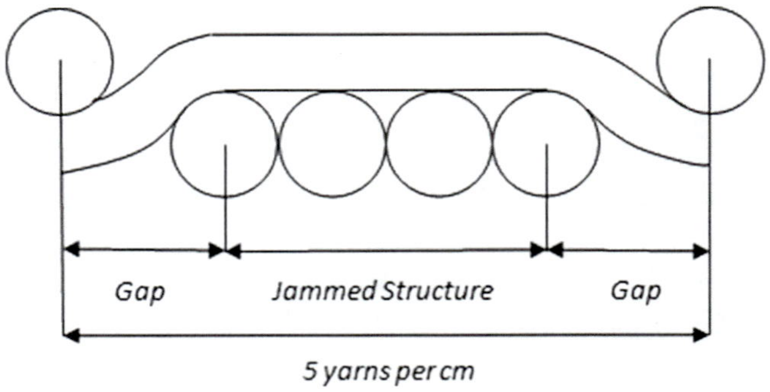

Fig. 4 Weave cross section of Basket 4/4 showing jammed structure

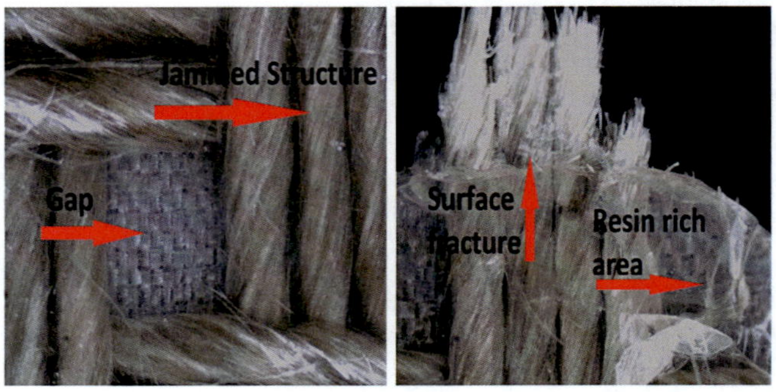

Fig. 5 Microscopy image of Basket 4/4 composite before and after tensile test

while the end point of interlacement generates a big gap without any reinforcement (Figs. 4 and 5). The jammed structure has great influence on the properties of the dry fabric such as the porosity where it affects resin penetration between fibers.

Figure 5 shows microscopic images of the structure and the fracture surface of basket 4/4 composite where the tensile fracture is shown as a crack on the resin-rich area and there is low resin penetration on the jammed structure. This has an influence on the composite tensile strength where the resin-rich area is the weakest point without any reinforcement materials, while low resin penetration on the fibers makes it low in strength as the resin is only on the surface of the yarns.

This is different with twill 4/4 and satin 8/3 where each adjacent floating yarn has a different interlacing point making it evenly distributed. Figure 6 shows an example of composite structure for satin 8/3 and the breaking fracture during tensile test. The structure of satin 8/3 shows even yarn floating distribution with small gap making it free for the resin to distribute evenly. The result shows a good fracture surface during tensile without any cracking from the resin-rich areas.

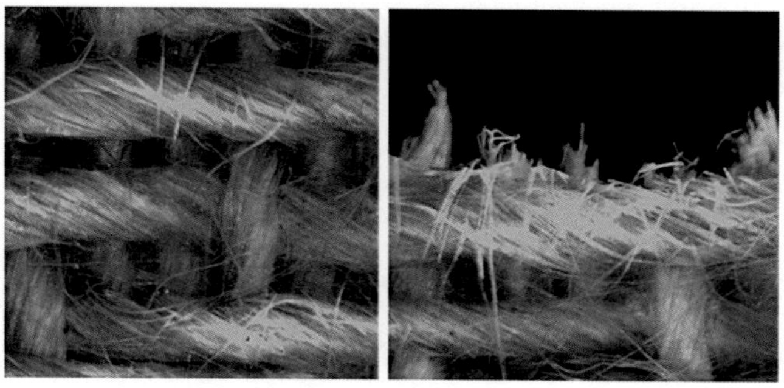

Fig. 6 Microscopy image of Satin 8/3 composite before and after tensile test

Conclusion

In this study, the effect of selected weave design increased the tensile strength of the woven kenaf reinforced unsaturated composite. The selection of the weave design is important as it has a great influence on the tensile properties of the dry fabric and composite. A weave design with low crimp percentage will increase the tensile strength of dry fabric by fully utilizing the fiber strength. However, the formation of the fabric structure may affect the tensile strength when it is in composite form. Some of the woven designs need attention during fabrication as the structure may change due to factors such as tension and contraction. It may not affect the properties of dry fabric but it may affect in composite form.

Acknowledgment The authors are grateful to the Faculty of Mechanical and Manufacturing Engineering and Advanced Textile Training Centre, Universiti Tun Hussein Onn Malaysia; Faculty of Chemical Engineering and Centre for Composite, Universiti Teknologi Malaysia; Department of Mechanical Engineering, Politeknik Seberang Perai; and Department of Metallurgy and Materials Engineering, Katholieke Universiteit Leuven for the WiseTex software.

References

1. A.K. Mohanty, M. Misra, G. Hinrichsen, Biofibres, biodegradable polymers and biocomposites: an overview. Macromol. Res. **276–277**, 1–24 (2000)
2. A. Causa, A. Netravali, Characterization and measurement of textile fabric properties, in *Structure and Mechanics of Textile Fiber Assemblies*, ed. by P. Schwartz (Woodhead Publishing Limited, Cambridge, 2008)
3. M.H. Peerzada, S.A.A.A. Khatri, Effect of weave structure on tensile strength and yarn crimp of three-dimensional fibre woven fabric. Sci. Int. (Lahore) **24**, 47–50 (2012)
4. P. Vandeurzen, J. Ivens, I. Verpoest, A three-dimensional micromechanical analysis of woven fabric composites: 1. Geometric analysis. Compos. Sci. Technol. **56**, 1303–1315 (1996)
5. Ž. Rukuižienė, R. Milašius, Influence of reed on fabric inequality in width. Fibres Text. East. Eur. **14**, 44–47 (2006)
6. B.K. Behera, J.M.R. Mishra, D. Kremenakova, Modeling of woven fabrics geometry and properties, in *Woven Fabrics*, ed. by H.Y. Jeon (InTech, Croatia, 2012), pp. 1–32
7. L.B. Manfredi, E.S. Rodríguez, M. Wladyka-Przybylak, Thermal degradation and fire resistance of unsaturated polyester, modified acrylic resins and their composites with natural fibres. Polym. Degrad. Stab. **91**, 255–261 (2006)
8. A.A.A. Rashdi, S.M. Sapuan, M.M.H.M. Ahmad, A. Khalina, Water absorption and tensile properties of soil buried kenaf fibre reinforced unsaturated polyester composites (KFRUPC). J. Food Agric. Environ. **7**, 908–911 (2009)
9. S. Lomov, I. Verpoest, F. Robitaille, Manufacturing and internal geometry of textiles in design and manufacturing of textile composite, in *Design and Manufacture of Textile Composite*, ed. by A. C. Long (Woodhead Publishing Limited, Cambridge, 2005), pp. 1–61
10. H. Reinforcements, *Technical Fabrics Handbook* (Hexcel Reinforcement for Composites, Texas)
11. T. Ngai, Carbon-carbon composite, in *Handbook of Composite Reinforcements*, ed. by S.M. Lee (Wiley, New York, 1993), pp. 41–70
12. J. Hu, *Structure and Mechanics of Woven Fabrics* (Woodhead Publishing Limited, Cambridge, 2004)

Yarn Pull-Out and Shear Behaviour of Kevlar 29 Fabrics Coated with Natural Rubber Latex

N.A. Ahmad, M.R. Ahmad, and W.Y.W. Ahmad

Abstract

This paper presents the yarn pull-out strength and shear behaviour of Kevlar fabrics coated with natural rubber latex (NRL). Plain woven Kevlar 29 fabric was coated with NRL using several dipping techniques which were single-dip (SD), double-dip (DD) and triple-dip (TD). The yarn pull-out and shear deformation were measured and observations were made on their modes of failure. The results indicated that the NRL-coated fabric showed higher load forces in comparison with the uncoated fabric. The yarn pull-out strengths for SD-, DD- and TD-coated fabrics were 17, 25 and 35 %, respectively, higher than the uncoated fabric. Similarly, the shear behaviour of the NRL-coated fabric also showed higher shear angles in comparisons with the uncoated fabric. It can be said that the NRL restricts the yarn movement, bunches them together to resist the external work force of yarn pulling or shearing before they completely failed at certain point. The presence of NRL layer on the fabric surface helps to increase the force loads and thus increase the yarn–yarn and yarn–NRL friction.

Keywords

Kevlar • Natural rubber latex • Shear behaviour • Yarn pull-out • Coated fabric

Introduction

Coated fabrics are currently used in many technical applications such as protective clothing, agricultural, architectural, automotive, geotextiles and medical [1–4]. The fabrics are normally coated with semi-liquid material such as rubber on one or both sides followed by a curing process which hardens the coating layer so that a non-blocking product is produced [1–5]. The application of viscous liquid can alter the properties and performance of the fabric. Parameters such as the type of fabric, coating elements, thicknesses and coating methods must be given careful consideration in order to achieve desirable end-use fabric properties [1, 2, 4]. These parameters also affect the properties of coated fabrics. Tests such as tensile strength, tearing strength, puncture strength, yarn pull-out and shear behaviour test are important to determine the qualities of the coated fabrics. In particular, yarn pull-out strength is an important indicator of yarn interactions within the fabric and as a prediction for other mechanical properties [6]. The test determines the strength of coating by pulling several yarns out from the fabric system without breaking the yarns. The test also gives the relative measure of yarn–yarn and yarn–NRL friction strength in the coated fabric [7].

There are a number of related researches concentrating on yarn pull-out strength [6–10]. Pan and Yoon [6] studied the behaviour of yarn pull-out of woven fabric with different types of weave and fibre types. They concluded that the maximum pull-out load is related to yarn geometric and mechanical properties and corresponding fabric count. Besides, different fabric weave structures and yarn types also give different yarn pull-out behaviours. Ahmad [7] reported that the effect of natural rubber latex (NRL) coating

N.A. Ahmad • M.R. Ahmad (✉) • W.Y.W. Ahmad
Faculty of Applied Sciences, Textile Research Center, Universiti Teknologi MARA, 40450 Shah Alam, Selangor Darul Ehsan, Malaysia
e-mail: rozitex@salam.uitm.edu.my

M.R. Ahmad and M.F. Yahya (eds.), *Proceedings of the International Colloquium in Textile Engineering, Fashion, Apparel and Design 2014 (ICTEFAD 2014)*, DOI 10.1007/978-981-287-011-7_6,

on yarn pull-out force was similar to that obtained from the tensile, tear and puncture strengths. In addition, the yarn pull-out force which increases after fabric coating indicates that the yarn–yarn friction in the fabric increases. Another study by Kirkwood et al. [8] reported that the length of the sample, the transverse fabric tension, the number of yarns pulled at once, the pattern of pull and the distance of yarn being pulled are dependent factors that affect the energy to pull a yarn out from a woven fabric. Sun et al. [9] investigated the yarn pull-out behaviour of five styles of plain woven Kevlar that differ in fabric count, yarn size, fibre type, thickness and weight. They found out that the single yarn pull-out test showed substantial difference in pull-out forces for the five styles of fabrics. All of these studies did not report the effect of coating techniques on yarn pull-out. Thus, an understanding of yarn pull-out strength for NRL-coated fabric is needed for various coating depths of techniques.

Apart from yarn pull-out, fabric shear is another important property of coated fabrics. Fabric shear deformation is very common during use since the fabric needs to be stretched or sheared to a greater or lesser degree as the body moves around [11–13]. Shear deformation or shear behaviour has generated some interest among textile researchers [11–17]. There are few methods that can be used to measure shear behaviour such as using the Kawabata evaluation system for fabric (KES-F), fabric assurance by simple testing (FAST) method or picture frame test (Fig. 1) [12]. The most common and simple test is the picture frame test or trellis frame because it can generate pure shear response within the fabric. Shear mechanism is one of the important properties influencing the draping, pliability and handle qualities of woven fabrics. It also can affect the bending, tensile and tearing strength of woven fabrics in various directions rather than in warp and weft directions only [11].

A study by Farboodmanesh et al. [12, 13] using polyester fabric coated with neoprene latex identified that the load values for rubber-coated fabric is higher in comparison to neat woven polyester fabrics. Their results also suggest that the shear behaviour of the rubber-coated fabric at low shear angles is dominated by rubber and at higher shear angles dominated by the fabric. Meanwhile, Nguyen et al. [14] reported experimental and theoretical shear properties of woven carbon fabrics and pre-pregs. The results indicated that the slip model gives modestly accurate predictions, whilst the elastic modulus model showed very good correlation with experimental data.

Another study by Naik [15] on Kevlar and Zylon fabrics showed a typical shear response where at the initial phase, the yarns begin to rotate, offering a small resistance to the applied shear loading. As the shearing continues, the fabric load tends to increase rapidly as the yarns compress each other laterally at crossover points. Galliot and Luchsinger [16, 17] proposed a new test method to investigate the shear behaviour of coated fabric. This method relies on the biaxial pre-tensioning of the orthotropic specimen where the fibres are aligned with the loading directions. This new method found that based on experimental observations and finite element analyses, the test method gives similar results for a wide range of coated fabrics if proper stress correction factors are used.

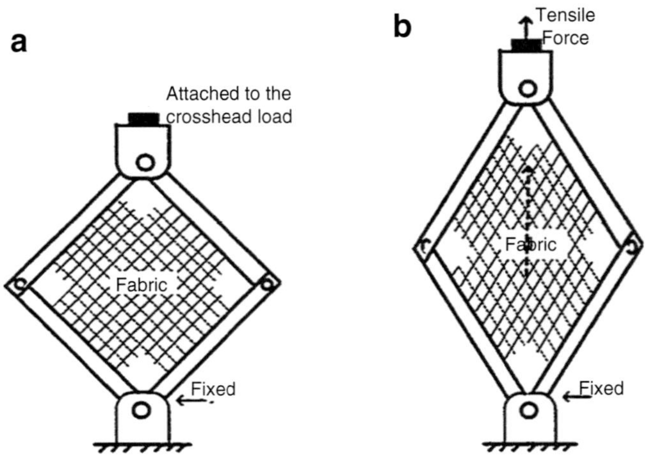

Fig. 1 Picture frame test (**a**) before shear and (**b**) after shear [12]

The objective of this study was to investigate the yarn pull-out strength and shear behaviour of the NRL-coated fabric and how the tests are related with the modes of deformation.

Materials and Experiment

Materials

Plain woven Kevlar 29 fabric was used in the study. The properties of the fabric are given in Table 1. Prior to coating, the fabric was first washed in order to remove all the lubricants or other foreign substances on the fabric surface. It was also meant to increase the bonding of NRL on the fabric's surface. Pre-vulcanized NRL was used as the coating material and supplied by Revertex Malaysia Sdn. Bhd. The properties of the NRL are shown in Table 2.

Coating Methods

Single-dip (SD), double-dip (DD) and triple-dip (TD) coatings were used to coat the Kevlar 29 fabric. Each sample was dipped in the NRL for 3–4 min to allow the NRL to adhere on the fabric's surface and then hang-dried at room

Table 1 Kelvar 29 fabric parameters

Parameters	Kevlar 29
Areal density	280 g/m^2
Yarn count	1,100 dTex
Warp and weft density	12 per cm
Thickness	0.419 mm

Table 2 Properties of NRL

Properties	NRL
Total solid content (%)	60.5
Viscosity (s), (Ford cup 3 at 25°)	30.5 s
Mechanical stability (s)	975
Tensile strength (MPa)	25.0

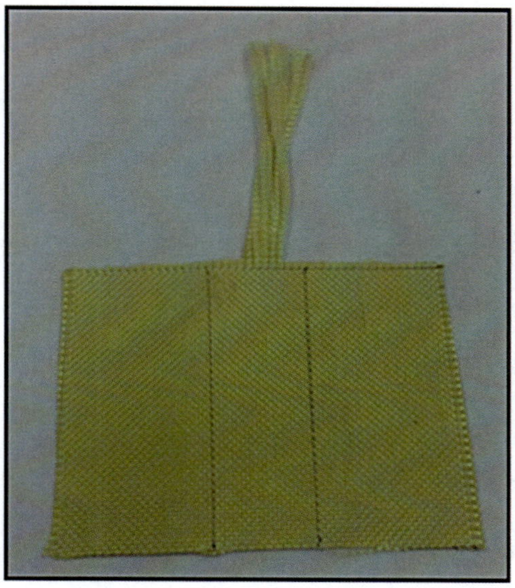

Fig. 2 Sample for yarn pull-out test

temperature. For the DD samples, a second dipping was done after 1 h or after the first layer of NRL has completely dried. The same goes for the subsequent TD coating. All the samples were finally left to dry at room temperature overnight.

Yarn Pull-Out Strength

Fabric samples of 80 mm × 120 mm were prepared with ten yarns at one end protruding at a certain length (Figs. 2 and 3). The yarns were pulled out from the fabric system without breaking it using the Testometric Strength Tester with 20 kN at a cross-head speed of 50 mm/min. The yarns were gripped using a modified clamp which has a 20 mm rectangle groove at the centre of the fabricated bottom clamp. The highest peak load (N) to start pulling the yarns out from the fabric sample was recorded, and an average of five readings was taken.

Fig. 3 Yarn pull-out strength test

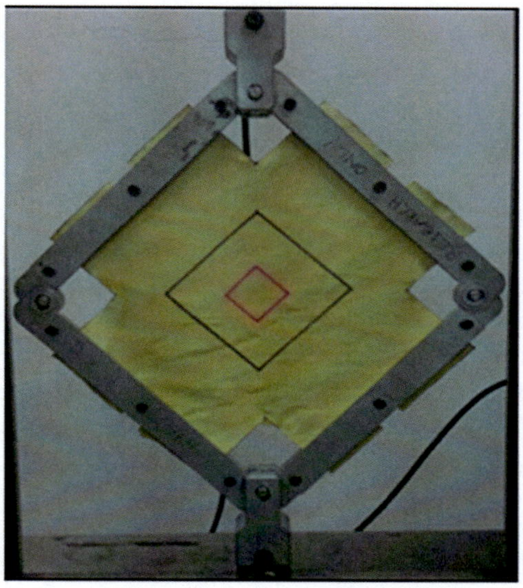

Fig. 4 Shear behaviour test

Shear Behaviour

A picture frame set-up as shown in Fig. 4 was used to test the shear behaviour of the samples as it gives pure shear response towards the fabric sample. A 280 mm × 280 mm fabric sample (Fig. 5) was clamped onto a fabricated clamp which has a groove on its surface to minimize slippage during experiments. The yarns were aligned in parallel and perpendicular to the sides of the frame test to reduce any additional tension in the yarns. The fixture was mounted vertically on the Testometric Strength Tester of 20 kN using a cross-head speed of 20 mm/min.

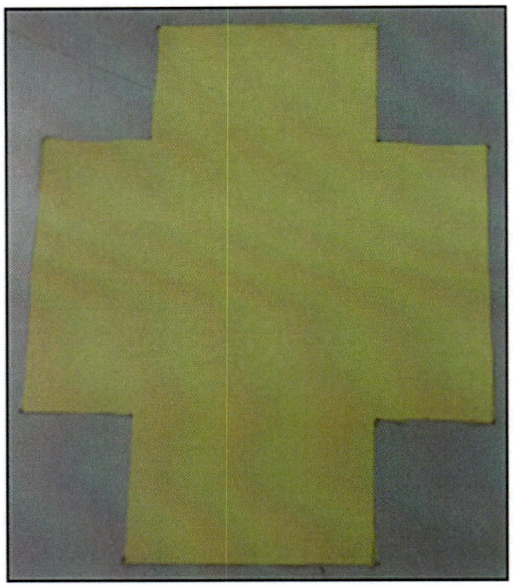

Fig. 5 Sample for shear behaviour test

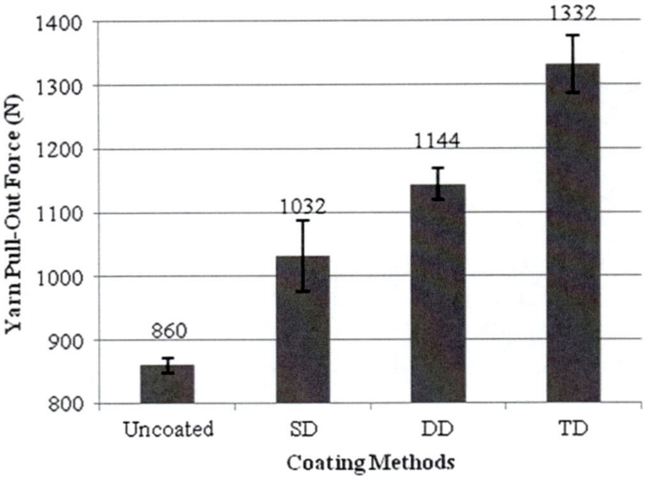

Fig. 6 Yarn pull-out force for different coating methods

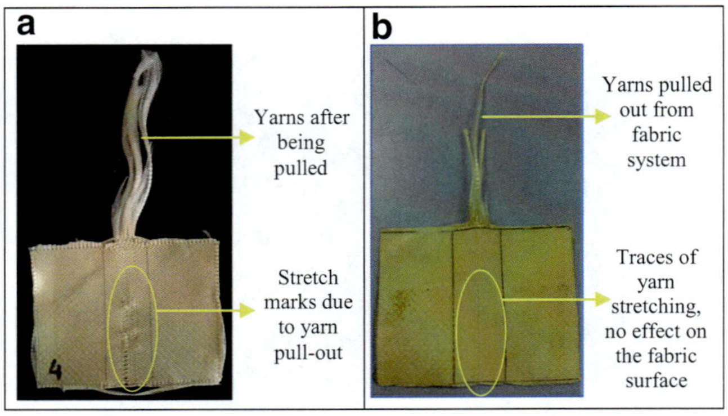

Fig. 7 Samples condition after yarn pull-out test. (**a**) Uncoated fabric sample. (**b**) NRL-coated fabric sample

Results and Discussions

Yarn Pull-Out Strength

The yarn pull-out strength was done to evaluate the relative measure of friction strength among the yarns in the fabric. Figure 6 clearly shows that all coated fabrics gave higher yarn pull-out strength compared to the uncoated fabric. The force loads for the SD-, DD- and TD-coated fabrics were higher than the uncoated fabric by 17, 25 and 35 %, respectively. The results showed a significant difference between uncoated and NRL-coated fabrics.

The results indicated that the yarn–yarn and yarn–NRL friction in the fabric increases after the coating process. The presence of an NRL layer on the fabric surface helps to increase the yarn pull-out load force. For the NRL-coated fabrics, the yarns had to overcome the high yarn binding forces imparted by the coating layer. The NRL film binds the yarns together in the fabric, restricts the movement and forces the yarns working together rather than individually to resist the external work load force. As the number of coating increases, the yarn friction in the fabric also increases as the NRL layer that forms on the fabric surface is much thicker. The increment of the thickness of the NRL layer makes it hold the fabric even tighter. On the other hand, for the uncoated fabric, there are fewer restrictions; thus, the yarns easily slide past each other once the pulled yarns overcome crimping and static friction. Hence, a lower load force was needed to pull the yarn from the fabric system. A higher yarn pull-out force will lead to higher tensile and tearing strengths as the friction between yarn–yarn and yarn–NRL in the fabric system increases.

The modes of the yarn pull-out start when the yarns overcome the static friction and this process involved yarn straightening [7]. After that, the yarns start to overcome crimp and start to stretch and slide over the orthogonal yarns. The maximum load force was eventually reached after this. Simultaneously, the orthogonal yarns around the pulled yarns tended to move in the direction of the pulling forces. These yarns were closer together and will give higher resistance forces. For the uncoated fabric, the pull-out forces will reduce after reaching the maximum load. Meanwhile, for NRL-coated fabric, the yarns might break at the grip upon achieving maximum load. Figure 7 shows some analysis on the samples.

Both the uncoated and SNRL-coated fabrics achieved the highest peak loads at different displacements. Figure 8 shows the graph for yarn pull-out force against displacement. The uncoated fabric achieved the highest peak load after being displaced at about 6 mm. This is because the uncoated fabric has fewer restrictions and the yarns are able to slide past each other easily. However, the yarns in the NRL-coated fabrics need to elongate more and work

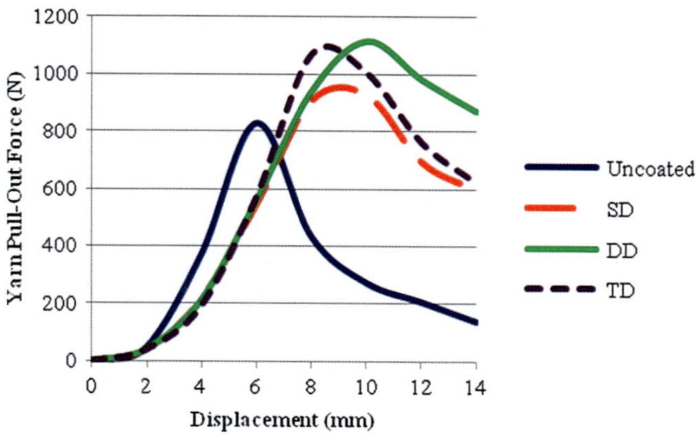

Fig. 8 Graph of yarn pull-out force (N) vs. displacement (mm)

together in order to be pulled out. Hence, they can achieve higher peak load force at higher displacement. The DD-coated fabric has the highest peak load of about 1,100 N and elongated more than 10 mm. The SD- and TD-coated fabrics were displaced somewhere in between the uncoated and DD-coated fabrics.

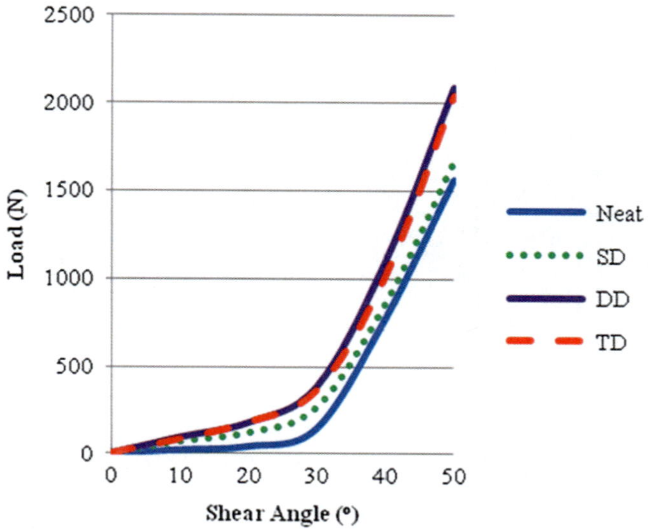

Fig. 9 Load vs. angle for shear behaviour

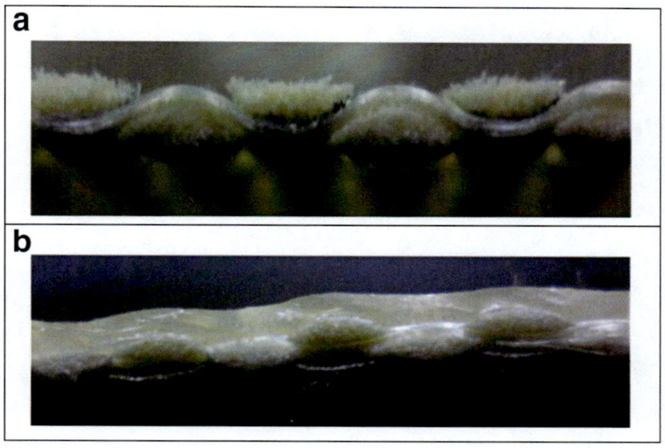

Fig. 10 Cross section of uncoated and NRL-coated fabric.(**a**) Cross section of uncoated woven Kevlar fabric. (**b**) Cross section of NRL-coated woven Kevlar fabric

Shear Behaviour

Figure 9 shows the result of the shear behaviour of uncoated and NRL-coated fabrics. It can be said that the lines follow the typical shear behaviour of a woven fabric. It can be clearly seen that the uncoated fabric has the lowest shear degree angle followed by SD- and TD-coated fabrics, whilst the DD-coated fabric has the highest shear degree angle. Hence, it can be said that the NRL layer on the fabric's surface helps increase the load force to shear the fabric.

In this study, only one type of fabric is used which is plain woven Kevlar fabric. Kevlar fabric is known to have a tight structure; thus, the degree of NRL penetration is lower and only able to form a layer on the fabric's surface. Thus, from the results, it can be said that the thickness of the NRL formed on the fabric's surface influenced the shear behaviour. In addition, as the number of the coating increases, the fabric rigidity also increases. Hence, when the fabric is being sheared, the yarns resist more and needed higher load force to resist the impact. Besides, the decreased yarn mobility for the NRL-coated fabric also increased the shear behaviour. Additionally, Fig. 10 shows the cross section of both uncoated and NRL-coated fabrics.

According to Farboodmanesh [12, 13], the shearing behaviour occurs in three stages: (1) an initial state where low shear loads are needed, (2) a yarn compaction region where the stiffness is increasing as the fibre volume fraction of the yarn increases and (3) a post-locking angle which is characterized by a sharp but relatively linear increase in load and if continued will cause fabric wrinkling. Figure 11 shows the shear behaviour stages from the initial stage (0°) until the last stage (>40°).

Figure 12 shows the shear load force against fabric elongation. It clearly indicated that the neat fabric has the lowest shear force, followed by SD- and TD-coated fabrics. The DD-coated fabric shows the highest shear force against elongation than that observed for TD-coated fabric. This could be due to some variations that may occur to the thickness of the fabric samples. However, the graph shows the same pattern at a slightly different peak for DD- and TD-coated fabrics. The

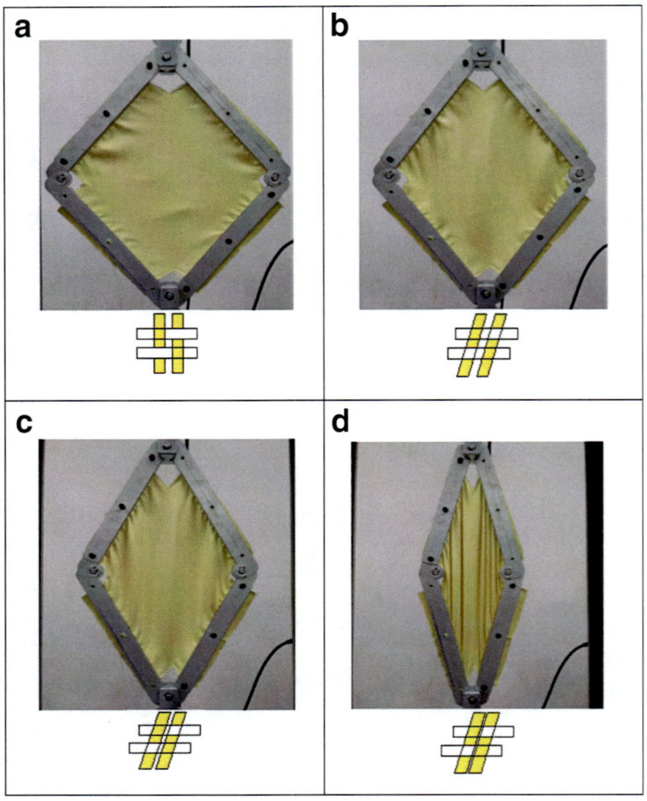

Fig. 11 Shear behaviour stages. (**a**) Initial state (0°). (**b**) Shearing (10°). (**c**) Compaction (20°–30°). (**d**) Locking angle (>40°)

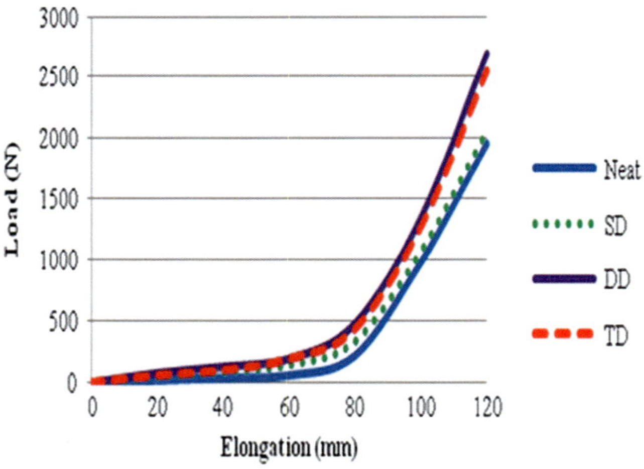

Fig. 12 Load vs. displacement for shear behaviour

NRL layer increased the shear resistance of the fabric and thus requires higher load force at about the same elongation point. It seems that the fabric thickness plays a role in determining the shear behaviour of both fabrics.

Conclusions

Overall, it was shown that the yarn pull-out and shear behaviour of NRL-coated fabrics were higher in comparison to the uncoated fabric. The stages that takes place in the yarn pull-out and shear behaviour were identified and in agreement with previous studies. The presence of an NRL layer on the fabric surface restricts the yarn mobility, bunching them together to resist external work load force which will then increase the force needed to completely fail them at a certain point of the test. This also proves that the increment of NRL coating layer increases the friction between yarn–yarn and yarn–NRL in the fabric system.

Acknowledgement The authors are grateful to Revertex (Malaysia) Sdn. Bhd. for supplying the pre-vulcanized natural rubber latex. The assistance from the Research Management Institute of Universiti Teknologi MARA is also highly appreciated.

References

1. A.K. Sen, *Coated Textiles: Principles and Applications*, 2nd edn. (Taylor and Francis, Boca Raton, FL, 2008)
2. W. Fung, *Coated and Laminated Textiles* (Woodhead Publishing Limited, Cambridge, 2002)
3. D.B. Wootton, *The Application of Textile in Rubber* (Rapra Technology Limited, Shawbury, 2001)
4. A.R. Horrocks, S.C. Anand, *Handbook of Technical Textiles* (CRC Press, Cambridge, 2000)
5. *Complete Textile Glossary* (Celanese Acetate LLC, New York, 2001)
6. N. Pan, M.Y. Yoon, Behaviour of yarn pull-out from woven fabrics: theoretical and experimental. Text. Res. J. **63**, 629–637 (1993)
7. M.R. Ahmad, Development of body armour system utilizing natural rubber coated fabric. Ph.D. Dissertation, Universiti Teknologi MARA, 2008
8. K.M. Kirkwood, J.E. Kirkwood, Y.S. Lee, R.G. Egres, E.D. Wtzel, N.J. Wagner, *Yarn Pull-Out As a Mechanism for Dissipation of Ballistic Impact Energy in Kevlar KM-2 Fabric, Part 1: Quasi-Static Characterization of Yarn Pull-Out* (Army Research Laboratory, 2004)
9. C.T. Sun, Z. Dong, Testing and modelling of yarn pull-out in plain woven Kevlar fabrics. Compos. Part A **40**, 1863–1869 (2009)
10. A. Gawandi, E.T. Thostenson, J.W. Gilllespie Jr., Tow pull-out behaviour of polymer-coated Kevlar fabric. J. Mater. Sci. **46**, 77–89 (2011)
11. J. Hu, *Structure and Mechanics of Woven Fabrics* (Woodhead Publishing in Textiles, Cambridge, 2004)
12. S. Farboodmanesh, Parametric studies of coated fabric shear behaviour: fabric construction and coating penetration effects. Master's Thesis, University of Massachusetts Lowell, MA, 2003
13. S. Farboodmanesh, J. Chen, J. Mead, K. White, Effect of construction on mechanical behaviour of fabric reinforced rubber. Rubb. Chem. Geometry **79**, 199–216 (2006)
14. M. Nguyen, I. Herszberg, R. Paton, The shear properties of woven carbon fabric. Compos. Struct. **47**, 767–779 (1999)
15. D. Naik, Experimental analysis of fabrics used in engine housing of aircrafts. Master's Thesis, Arizona State University, Phoenix, 2005
16. C. Galliot, R.H. Luchsinger, The shear ramp: a new test method for the investigation of coated fabric shear behaviour – Part I: theory. Compos. Part A **41**, 1743 (2010)
17. C. Galliot, R.H. Luchsinger, The shear ramp: a new test method for the investigation of coated fabric shear behaviour – Part II: experimental validation. Compos. Part A **41**, 1750 (2010)

Tensile Strength and Evenness of Kenaf/Polyester Blended Rotor-Spun Yarn

N.H.A. Hayam, M.R. Ahmad, W.Y.W. Ahmad, M.F. Yahya, and M.I.A. Kadir

Abstract

This paper reports on some properties of kenaf/polyester blended yarns which were spun using rotor spinning at different rotor speeds and fibre blends ratio. The blended yarns were spun at 50,000; 60,000; and 70,000 rpm rotor speeds. The kenaf and polyester fibres were mixed at the carding process at percentages of 5/95, 10/90 and 15/85, respectively. The size of the yarn was kept constant at 40 Tex. It was found that with higher percentage of kenaf fibres, the yarn tensile strength decreases at every rotor speed. The yarns produced from 5/95 blending ratio at rotor speed of 60,000 rpm have the highest strength among the other yarn samples. The yarn imperfections also showed similar trend as the tensile strength. The study also showed that the yarn evenness and imperfections were lowest for yarns with the least percentage of kenaf fibres.

Keywords

Kenaf fibres • Rotor spinning • Rotor-spun yarns • Yarn blending

Introduction

Kenaf is a cellulosic fibre that belongs to the same group as cotton. It is an annual crop that originated from Africa and has become an important crop in many countries for commercial market. The fibres are now grown abundantly in countries such as India, Bangladesh, Malaysia, Indonesia, the United States, Thailand and Vietnam [1]. This plant is an ancient crop similar to jute but has better lustrous and tensile strength [2]. The plant consists of two types of fibres which is the outer bark or bast and the inner woody core. The bast fibre constitutes 35 % of the core containing short fibres which makes 65 % of the plant [3]. The kenaf plant is unbranched and grows quickly with a height of 8–20 ft in 4–5 months [4].

Kenaf is a multipurpose plant with the entire kenaf plant, leaves, tender shoot, woody core, bast fibres and seed, valuable in different usages. The end products of kenaf fibres depend on the portion fibre used [1]. However, kenaf fibres are usually highly productive for their stalk from which the fibre is extracted. Traditionally, the allied fibres are used as cords, ropes and twine [5], and many studies have also shown that it offers good opportunities as reinforcement material for composites. Kenaf fibres also offer advantages as composites being lightweight and environmentally friendly. Yuhazri [6] studied the mechanical properties of kenaf/polyester and reported that kenaf/polyester processed by vacuum infusion can replace existing materials with better strength, low cost and environmental friendliness. Another study by Jeyanthi and Rani [7] reported that hybrids of kenaf-/glass-reinforced composites can be utilised for passenger car bumper beam. Juliana and Paridah [8] claimed that kenaf stem offers good advantage in making particle-board based on the high-density part of the stem. Apart from that, kenaf fibres also have the great advantage of providing pulp to the newsprint industry [9].

In Malaysia, kenaf plants are considered to be an alternative and versatile fibre crop as the fibres are completely biodegradable, renewable and environmentally friendly

N.H.A. Hayam • M.R. Ahmad (✉) • W.Y.W. Ahmad • M.F. Yahya • M.I.A. Kadir
Faculty of Applied Sciences, Textile Research Center, Universiti Teknologi MARA, 40450 Shah Alam, Selangor Darul Ehsan, Malaysia
e-mail: rozitex@salam.uitm.edu.my

M.R. Ahmad and M.F. Yahya (eds.), *Proceedings of the International Colloquium in Textile Engineering, Fashion, Apparel and Design 2014 (ICTEFAD 2014)*, DOI 10.1007/978-981-287-011-7_7,

which does not require a lot of chemicals in processing or cultivation [10]. However, only a few researchers investigated the development of converting kenaf fibres into yarn.

One of the downstream processes is to spin the fibres into yarns. This has typically been a challenge since 100 % kenaf fibres are extremely difficult to spin using any of the modern spinning systems available. This is due to characteristics of kenaf fibres which are somewhat brittle and coarse. One alternative is to blend the fibres with other conventional fibres such as polyester and spin them using rotor spinning. The combination of cotton fibres with polyester is common, and these fibres have been the benchmark. This paper presents some studies in the spinning of kenaf/polyester blends and the effects of rotor speeds on the tensile strength, evenness and imperfection properties of the yarn.

Fig. 1 Kenaf fibres ready for mixing with polyester

Materials

Sliver Preparation

The kenaf fibres were supplied by a local kenaf plant provider, Kenaf Natural Fiber Industries Sdn. Bhd., Pasir Puteh, Kelantan, Malaysia. The fibres were extracted manually after retting them in water for a few days and dried under the sun. The fibres were cut according to the specified average length of 40 mm (Fig. 1).

The sliver formation was conducted at a local spinning mill, Recron (Malaysia) Sdn. Bhd., Nilai, Negeri Sembilan. The kenaf and polyester fibres were prepared into three different ratios: 5/95, 10/90 and 15/85. The polyester fibres were 1.2 denier in size with an average length of 32 mm and were mixed with kenaf fibres by weight basis based on the percentages mentioned earlier. After mixing, they were manually fed into the feed roller of the carding machine. Here, the kenaf fibres were opened into individual fibres and blended with the polyester based on the ratio mentioned earlier. The carding process basically improved fibre evenness and parallelization and converted the fibre blends into a sliver form with an average size of 6,700 tex.

The kenaf/polyester blended slivers were made to pass through the first and second drawframe passages before they were finally evenly blended together.

Yarn Preparation

The slivers from the 2nd passage drawing were spun using the single-rotor unit of the SDL Atlas Quickspin rotor spinning machine (Fig. 2). They were spun using three different rotor speeds: 50,000; 60,000; and 70,000 rpm.

Fig. 2 SDL Atlas Quickspin rotor spinning machine

Other parameters kept constant were the rotor diameter (40 mm), opening roller speed (6,000 rpm) and delivery speed 75 m/min. The kenaf/polyester blended rotor-spun yarns were labelled C1 to C9 as shown in Table 1.

Methodology

Yarn Tensile Properties

The single yarn tenacity and elongation were tested using Uster Tensojet (UTJ) with a gauge speed of 100 m/min and a test length of 50 cm. An average of 20 readings for the tenacity and elongation were taken.

Table 1 Kenaf/polyester rotor-spun yarn samples

Sample no.	Rotor speed (rpm)	Blends ratio (%)
C1	50,000	5/95
C2	50,000	10/90
C3	50,000	15/85
C4	60,000	5/95
C5	60,000	10/90
C6	60,000	15/85
C7	70,000	5/95
C8	70,000	10/90
C9	70,000	15/85

Yarn Evenness

The yarn evenness (CV%) and imperfections (thick places, thin places and neps) were measured using Uster Tester 3 with a speed of 25 m/min. The degree of sensitivities were −50, +50 and +280 %, respectively.

Microscopic View

The yarn longitudinal structure and wrapper fibre formation of the blended yarns were observed under the Pro10 Digital Mobile Microscope and Scanning Electron Microscope. They were done to observe the positions of the kenaf fibres within the yarn structure.

Table 2 Yarn tenacity and elongation

Rotor speed (rpm)	Yarn sample no.	Tenacity (cN/Tex)	Elongation (%)
50,000	C1	19.49	9.03
	C2	19.20	9.17
	C3	19.60	9.28
60,000	C4	20.96	8.95
	C5	20.25	9.05
	C6	18.13	8.31
70,000	C7	18.57	9.46
	C8	17.57	8.00
	C9	17.15	7.81

Table 3 Test result for yarn irregularities and imperfections

Samples	CVm (%)	Thin (−50 %)	Thick (+50 %)	Neps (280 %)	Hairs/m
C1	17.15	27	800	347	5.38
C2	18.26	0	1,213	600	5.42
C3	17.64	0	1,000	320	5.44
C4	16.03	0	867	560	5.19
C5	19.11	13	1,547	880	5.16
C6	19.83	0	1,720	1,080	5.47
C7	17.36	0	1,027	587	6.01
C8	19.80	13	1,413	960	6.53
C9	20.21	0	1,493	1,120	6.53

Results and Discussions

Yarn Tenacity and Elongation

The effects of different rotor speeds on the kenaf/polyester blended rotor-spun yarns are shown in Table 2. It can be seen that at rotor speeds of 50,000 and 60,000 rpm, the yarn tenacity did not show any trend as the results were similar. However, at a rotor speed of 70,000 rpm, the yarn tenacity decreases. The elongation at break of the samples did not show any significant trend and can be said as more or less the same for all rotor speeds.

The rotor speeds used in the study are in the low- to medium-speed category and did not appear to give significant differences in terms of tenacity and elongation regardless of the percentages of kenaf used. Spinning of the yarns at much higher rotor speeds is proposed for future works.

Yarn Evenness

The thin places, thick places and neps are important for the determination of yarn quality and will affect yarn appearance and yarn irregularities. Table 3 shows the relation between the irregularities and imperfections with the different types of kenaf/polyester blended rotor-spun yarns. The results suggest that at every rotor speed, the yarns with the most kenaf percentage gave a higher value of CV%. A high value of CV% indicates high yarn unevenness or irregularity [11]. The results also suggest that at higher rotor speeds, the yarn irregularity increases. The trend was similar with thick places and neps.

Microscopic View

The general structure of rotor-spun yarns consists of two parts: the core characterised by helical twisting of fibres and the sheath constituted by individual or thin ribbons of wrapper fibres which form belts on the yarn surface. Figures 3 and 4 show the structure of the kenaf/polyester blended yarn under light microscope, while Figs. 5 and 6 show the structure under SEM.

It can be seen that in general, the polyester fibres are dominant in the yarn core and as wrapper fibres. This is expected since the majority of the fibres are polyester. Kenaf fibres did not show to become wrapper fibres but are seen lying within the yarn core and surface. Figures 5 and 6 suggest that there are loose and tight wrapper fibres within the yarn structure.

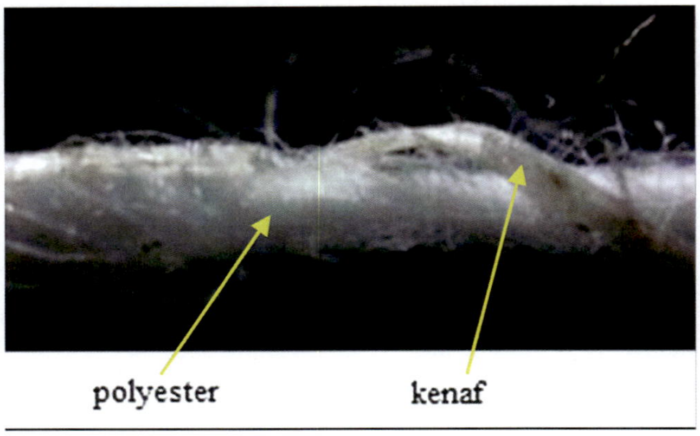

Fig. 3 The position of kenaf and polyester in the yarn structure

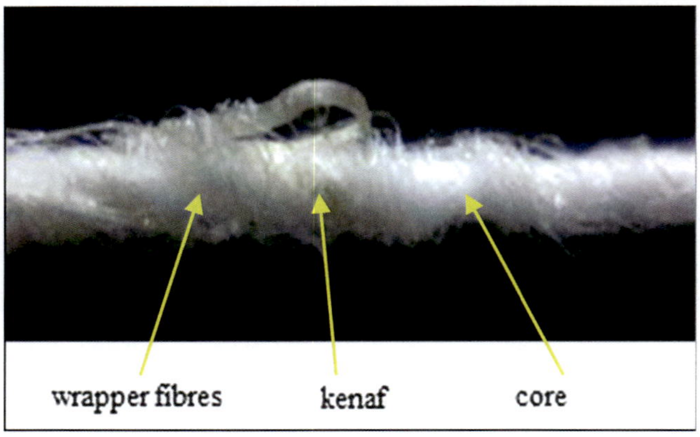

Fig. 4 The structure of the kenaf/polyester blended yarn

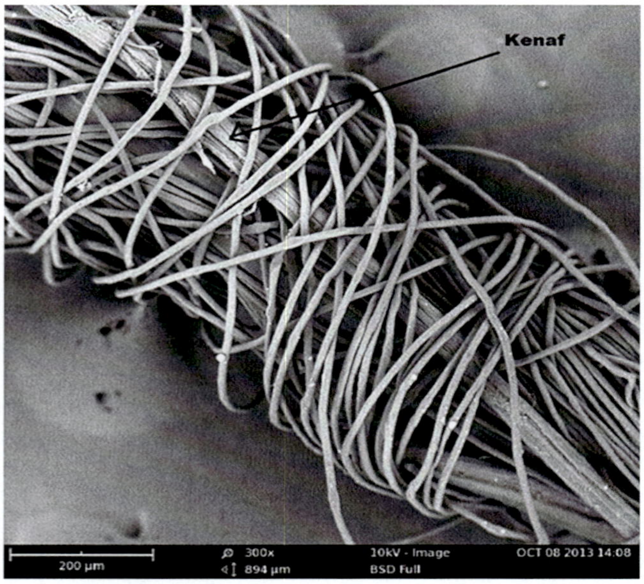

Fig. 5 The structure of the kenaf/polyester blended yarn under SEM

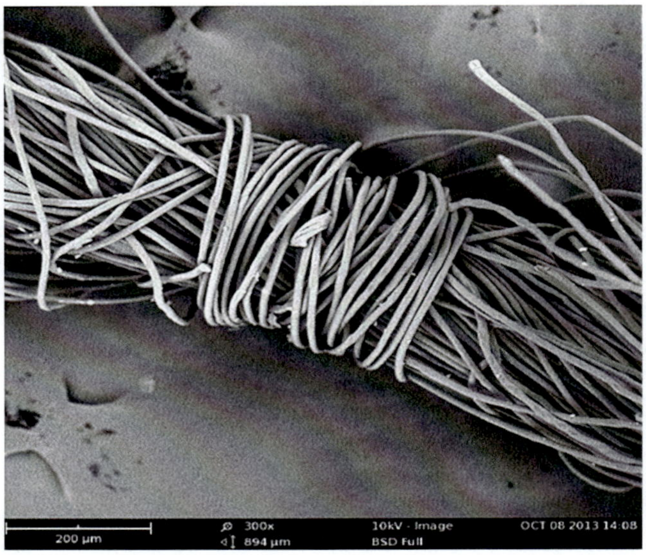

Fig. 6 The structure of the kenaf/polyester blended yarn showing tight wrapper fibres

Conclusions

The different percentages of kenaf and polyester fibres in the yarn have some effects on the yarn quality in terms of yarn strength and imperfections. The rotor speeds used in the study gave slightly different results for yarn evenness, thick places and neps, but did not show significance differences for thin places and elongation. The kenaf fibres also did not show to become wrapper fibres but rather lie within the yarn core. Overall, yarns with 5 % kenaf gave higher tensile strength, yarn evenness and imperfections for every rotor speeds. The effects of higher rotor speeds and other spinning parameters are proposed for future studies.

Acknowledgement The authors would like to thank Recron (Malaysia) Sdn. Bhd for assisting in the sliver preparation and yarn testing. The project is funded under eScienceFund from the Ministry of Science, Technology and Innovation, Malaysia.

References

1. T. Sen, H.J. Reddy, Various industrial applications of hemp, kinaf, flax and ramie natural fibres. Int. J. Innov. **2**(3), 192 (2011)
2. T. Zhang, Improvement of Kenaf yarn for apparel applications. University of Chemical Technology, 2003
3. C.L. Webber, J. Whitworth, J. Dole, P.O. Box, H. West, Kenaf (*Hibiscus cannabinus* L.) core as a containerized growth medium component. Ind. Crop. Prod. **10**, 97–105 (1999)
4. Y.A. El-Shekeil, Influence of chemical treatment on the tensile properties of kenaf fiber reinforced thermoplastic polyurethane composite. Express Polym. Lett. **6**(12), 1032–1040 (2012)

5. C.L. Webber, V.K. Bledsoe, Plant maturity and kenaf yield components. Ind. Crops Prod. **16**(2), 81–88 (2002)
6. M. Yuhazri, H. Sihombing, A.R. Jeefferie, K. Rassiah, Mechanical properties of Kenaf/polyester composites. 106–110 (2011)
7. S. Jeyanthi, J.J. Rani, Improving mechanical properties by kenaf natural long fiber reinforced composite for automotive structures. J. Appl. Sci. Eng. **15**(3), 275–280 (2012)
8. A.H. Juliana, M.T. Paridah, Evaluation of basic properties of Kenaf (*Hibiscus Cannabinus* L.) particles as raw material for particleboard. in *18th International Conference on Composite Materials*, vol. 36, pp. 1–6
9. P. Sullivan, Kenaf production (2002) [Online]. Available: http://www.attra.org/attra-pub/PDF/kenaf.pdf. Accessed 10 Oct 2013
10. I. Fulton, M.S. Qatu, Mechanical properties of Kenaf-based natural fiber composites. in *16th International Conference on Composite Structures*, 2011
11. N.A. Kotb, Predicting yarn quality performance based on fiber types and yarn structure. Life Sci. J. **9**(3), 1009–1015 (2012)

SMART Textiles: The Use of Embedded Technology on Tactile Textiles as Therapy for the Elderly

K. Hong

Abstract

This research project focuses on the therapeutic abilities of Tactile Textiles for the use of the elderly in Singapore as a form of touch therapy, especially for those with deteriorating senses. The main aim of the project is to use the technique of heat setting to form Tactile Textiles to stimulate the sense of touch. With the use of embedded technology, a series of textile-based cushions with tactile surfaces responding to the patient's sense of touch and sight have been developed and tested by the elderly in an eldercare centre.

Keywords

Heat setting • Three-dimensional textiles • SMART embedded technology • Deteriorating sense of touch • Elderly

Introduction

Touch is a very important sense. Through our skin, we receive information regarding touch, pressure, texture, temperature and pain. Being able to react positively to touch sensations enables us to feel comfortable and emotionally secure. However, there are certain groups of people who will have difficulty in accepting touch or are experiencing a decline in the sense of touch. To help with the well-being of these selected groups of end users, a series of products will be designed specifically for their usage as therapy to enhance, improve and encourage the sense of touch.

Tactile Textiles is a series of textile-based materials and products that incorporates SMART technology to produce 'cause and effect' upon touching of the products. The textiles will be aesthetically designed and constructed as three-dimensional forms by utilizing the thermoplastic qualities of synthetic fabrics during the first stage. The design of the three-dimensional forms will attract the users to feel and touch the textiles, and with the seamless incorporation of SMART technology such as motion sensors, which will detect hand movements on the surface of the fabrics, sensory outputs such as light and heat will be produced.

The main target groups of end users are the elderly patients who are suffering from a deteriorating sense of touch. With the use of embedded technology and using textiles as a base material, a series of textile-based products with 3D tactile surfaces responding to the patient's sense of touch and sight have been developed. As one of the common consequences of aging is a decline in the sense of touch, the elderly will be able to practise their tactile responsiveness on the designed 3D tactile fabric, hence becoming a necessary therapy in enhancing their tactile sensation. The use of embedded technology incorporated into these textiles will allow the elderly to experience both tactile and visual sensory at once.

Tactile Therapy

One of the common consequences of aging is a decline in the sense of touch. Schmall [1] further elaborates that touch sensitivity decreases with age. Some older persons find it difficult to distinguish textures and objects on the basis of

K. Hong (✉)
Singapore Institute of Technology, Singapore, Singapore
e-mail: karen.hong@singaporetech.edu.sg

M.R. Ahmad and M.F. Yahya (eds.), *Proceedings of the International Colloquium in Textile Engineering, Fashion, Apparel and Design 2014 (ICTEFAD 2014)*, DOI 10.1007/978-981-287-011-7_8,

touch alone. Some may experience a delayed reaction to being touched. To prevent further deterioration of the touch sensitivity, the design of the Tactile Textiles with its distinct textures created from three-dimensional structural forms will provide haptic sensations during the therapeutic interventions.

With the focus in stimulating the sense of touch for this particular group of end users, the elderly will be able to practise their tactile responsiveness on the three-dimensional tactile fabrics, hence becoming a necessary therapy in enhancing their tactile sensation. The use of embedded technology incorporated into these textiles will help maintain touch sensitivity and at the same time allow them to enjoy interactive play or provide comfort.

For example, an elderly in a nursing home could be covered with a tactile quilt or hug a tactile cushion with a three-dimensional structural form created from textiles. Light will be triggered when the elderly touches or strokes the surface of the Tactile Textiles. This is made possible by attaching motion sensor systems linked with LED lights woven onto the base fabric of the material. Other possibilities can also include the emission of heat or even triggering a vibrating motor while the surface is being touched. The outcomes of light, heat or vibration will be beneficial to the sensory responsiveness of the elderly. These outcomes will be fun examples of cause-and-effect play for the elderly, which will stimulate all of the senses, and essentially lead to a more fulfilled, worthwhile and happier quality of life.

For the elderly, this research project focuses on the therapeutic abilities of Tactile Textiles for the use of the elderly in Singapore as a form of touch therapy, especially for those with deteriorating senses. The main aim of the project is to use the technique of heat setting to form Tactile Textiles to stimulate the sense of touch. With the use of embedded technology, a series of textile-based cushions and other products with tactile surfaces responding to the patient's sense of touch and sight have been developed and tested by the elderly in an eldercare centre.

There are different types of therapy in geriatric daycare and eldercare centres in Singapore. Some are enjoying sessions of art as therapy which includes simple basic Chinese paintings and easy-to-manage handicraft making. Hands-on activities like these allow the elderly to maintain their visual skills, sensory skills as well as motor skills.

It has been proven from research by the hospital that weekly arts and crafts sessions can reinforce motor skills and challenge the patient's creativity. Besides art therapy, simple materials and play objects are available for the patients to work on their motor skills. There are several boards with bolts fastened on and patients will practise their motor skills by screwing the lock nuts onto the bolts. However, these materials lack visual interest and interactivity to further stimulate the patients. Using this as a starting point, a series of visually stimulating and interactive set of materials can be designed for the patients.

There are ongoing projects in Europe that combine SMART materials within textiles. Swedish RE:FORM Design Studio [2] at Goteborg invented an interactive cushion. The cushions come in pairs; the idea is that when one is touched or squeezed, the other glows dynamically. The designers of these cushions believed that by hugging these interactive cushions, long-distance relationships can be more bearable.

It is also proven by scientific research that animal-assisted therapy [3] is a respected therapeutic intervention. Holding, petting and simple grooming tasks can enhance motor skills lost through injury, disease or the aging process. With this study, the use of the tactile qualities of fabrics is also similar. It will allow patients to utilize their sense of touch and further illustrates the therapeutic abilities of the project's hypothesis.

Heat Setting to Create Three-Dimensional Textiles

The integral part of Tactile Textiles that is the first stage will be the design and creation of the three-dimensional structural forms of textiles. These three-dimensional structural textiles are formed by using the heat-setting techniques which is a technical approach of fabric manipulation that provides a platform into some of the most creative and innovative approaches to surface and textile design. This technique enables a flat fabric to be transformed into structural and sculptural forms. The creative process of pleating, crushing as well as moulding continues to evolve into different possibilities, hence creating this range of interesting surface designs that is fundamental to the design process of this project.

The textiles will be aesthetically designed and constructed as three-dimensional forms by utilizing the thermoplastic qualities of synthetic fabrics. Polyester, with thermoplastic qualities in the form of fabrics and fibres, can be given a three-dimensional form regardless of construction methods and amalgamation with different surface design techniques. By definition, thermoplastic [4, pp. 177] refers to the quality of a fibre whose molecular structure breaks down and becomes fluid at a certain temperature, making it possible to reshape the fabric by pleating, moulding or crushing. The fabric is 'fixed' on cooling and cannot be altered unless heated to a temperature greater than the one at which it was reshaped.

Polyester belongs to the group of synthetic fibres. Some other examples of the synthetic fibre group include polyamide, acetate, acrylic, viscose and elastane. Polyester is thermoplastic, that is, it can be transformed through heat into new configurations, which on cooling are completely stable [5, pp. 72]. Polyester fibres and fabrics, being thermoplastics, can be given a new form by heat setting. In this project, different techniques of heat setting will be explored to show the flexibilities of polyester being given a three-dimensional structure and form.

Moulding of the fabric creates structural surfaces on the fabric. Moulds of different shapes and sizes can be used to provide the three-dimensional effect of the fabric. One easy approach is to apply simple 'shibori' techniques, by using a binding technique; glass marbles can be tied to the polyester organza. Then the fabric can be placed into the microwave or oven for the heating process. Once heated, the fabric will adhere to the form of the marble mould, which is below the melting point of the fabric. Once the fabric has been heat set, the marbles are removed. This creates an amazing textural, rounded and three-dimensional effect to the fabric. This structural design can be customized to one's liking; whether it is done closed together or further apart, it creates a different form with each individual fabric.

Shibori is a centuries-old traditional Japanese textile finishing technique. It involves the tying and folding of a fabric before the dyeing process. This technique is originally used on silks and plant fibres, which leads to unique patterns, textures, structural forms and colours after the dyeing process. On natural fibres, the three-dimensional effect will not be permanent. However, on synthetic fibres such as polyesters, when treated with the shibori technique and heat, the fabrics will result in a permanent three-dimensional surface.

The inspiration of the structural three-dimensional shape comes from natural and organic elements. Hence, a range of names is given to these different shapes that evoke the natural organic patterns of nature. There are five styles of fabric formations, namely 'meanders', 'floras', 'shrooms', 'stalacs' and 'bubbles'.

The 'meanders' are made from rows of repeated pleats, and when the user pulls the pleats apart, it will crinkle and form meandering patterns. 'Floras' are made from flat constructed round disc shapes, of varying sizes, which reminisce an oriental coin with a hole in the middle, known as 筒 (tóng). 'Shrooms' are made from flat constructed round disc shapes. 'Stalacs' are done by the shibori style of stitching and binding the fabrics into mini cone shapes. Lastly, 'bubbles' are also created by the binding method, by tying marbles onto the fabric.

These three-dimensional forms not only conjure the imagery of the natural elements but also intrigue and induce the user's sense of touch. These shapes, with the texture of the organza fabric, feel interesting to caress and massage over the user's skin (Fig. 1).

Fig. 1 The five styles of fabric formations, namely 'meanders', 'floras', 'shrooms', 'stalacs' and 'bubbles'

Embedding of SMART Technology

Subsequently, the next stage is the seamless incorporation of SMART embedded technology within Tactile Textiles. The scientific definition of 'SMART textiles' as defined by Stylios, Stylios [6] are materials that can sense mechanical, thermal, chemical, magnetic or other environmental conditions and respond in a controlled and predicted manner by changing their colour, permeability, porosity, rigidity, shape, size or other physical/chemical characteristics. As explained by Langenhove and Hertleer [7] the term 'SMART textiles' is derived from intelligent or SMART materials. The concept of 'SMART material' was, for the first time, defined in Japan in 1989. Generally, 'SMART textiles' are able to sense stimuli from the environment, to react to them and adapt to them by the integration of functionalities in the textile structure.

In this research project, motion sensors are embedded onto the Tactile Textiles. The stimuli in this case will be the touching or stroking of the surface of the Tactile Textiles. The reaction output will be light emitting out from the Tactile Textiles. This is made possible by attaching motion sensor systems linked with LED lights woven onto the base fabric of the material. The other possibility includes the emission of heat while the surface is being touched. The outcomes of light and heat will be used as a form of tactile touch therapy that could be integrated into current tactile treatments for the elderly.

The main application for this range of heat set fabrics will be to develop them into cushion covers. With its tactile qualities, it will encourage users to feel and stroke the cushions. With the use of embedded technology and using textiles as a base material, a series of textile-based equipment with tactile surfaces responding to the patient's sense of touch and sight has been developed. As one of the common consequences of aging is a decline in motoric

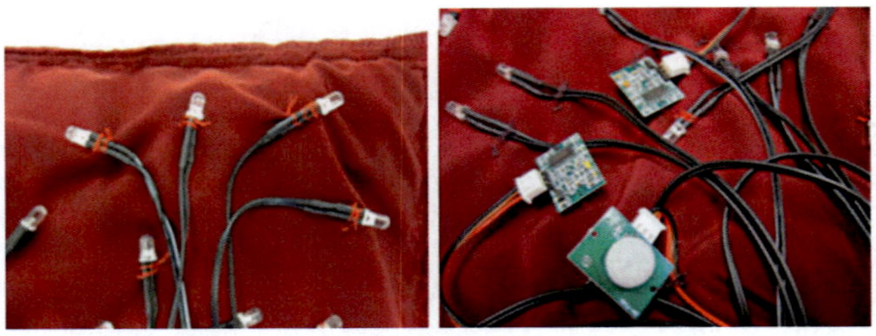

Fig. 2 Placing of custom-made LED with motion sensor on the inside of the cushion

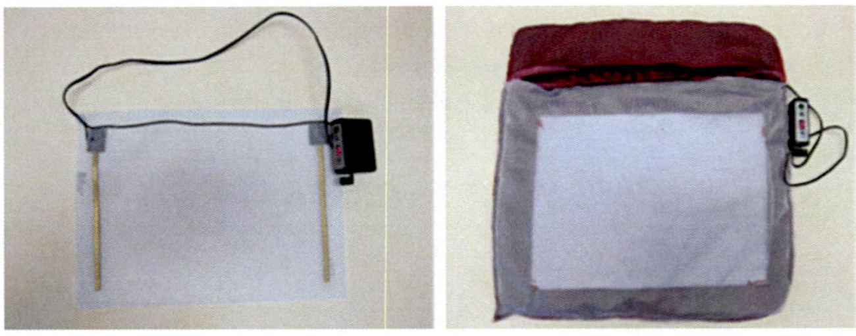

Fig. 3 Placing of SEFAR PowerHeat on the inside of the cushion

and sensory functions, patients or the elderly will be encouraged to feel and touch the tactile material to enhance their tactile sensory input. The addition of embedded technology incorporated into these textiles will allow patients or the elderly to experience different sensory inputs.

The use of embedded technology incorporated into these cushions will allow the elderly to enhance their sensory experience. The main objective will be to incorporate and weave both tactile qualities with the SMART technology seamlessly. With the outcomes of the experimentations, a series of visually stimulating and interactive set of cushions are designed for the elderly. When the tactile surfaces of the cushions are being stroked, cushions with LED lights embedded will be lighted up. This is made possible by having custom-made LED lights placed underneath the 3D surface with 4 motion sensors. The LED lights will be lighted up when the 3D surfaces are being stroked (Fig. 2).

Cushions are also being embedded with a heated material, which could provide warmth to the user, to be used in an air-conditioned room or in cold countries. The bases of the cushions are installed with a lightweight textile heating system produced by SEFAR. The product is known as SEFAR PowerHeat, and it is a light and highly air-permeable heating fabric [8]. The cushion will emit heat at the base, which will provide warmth to the patients when needed, and at the same time, patients can enjoy the different tactile sensations when they stroke the surfaces of the cushions (Fig. 3).

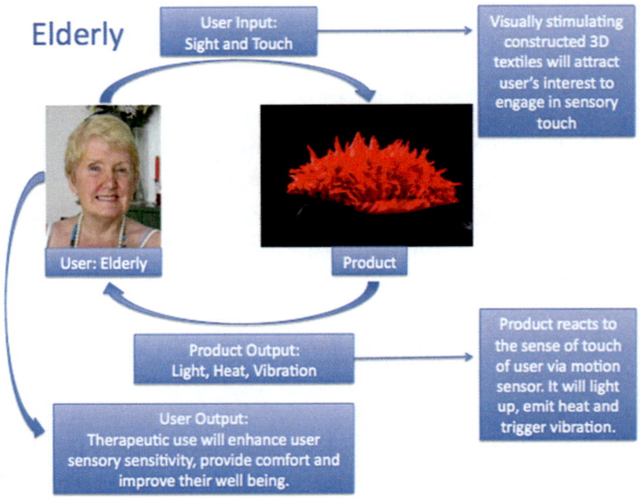

Fig. 4 Rationale for design

Results and Discussions

The heated cushions prove to be the most useful when tested on the elderly. Most of them commented that the heat not only provides them with warmth; they are able to ease their joint pains and rheumatism. Many of them wished to own a heated cushion and to use it during bedtime as the heat will be effectively soothing as most of them face joint pains at night.

The lighted cushions offer the elderly an interactive play of the 3D tactile surfaces. Many of them commented that they do like the 3D tactile qualities of the cushions. While stroking, it will evoke different sensations and allows them to play and stroke the tactile surfaces. The lighted LED gives them an element of surprise and fun to keep stroking the cushions.

These cushions will serve to allow the 3D tactile fabric to be integrated as much as possible into the everyday lives of the elderly, and in doing so, making the therapy more effective. They are designed in such a way that it will be touched and used as much as possible without having to be forced onto the elderly patients. The user experience is fun, interactive as well as effective to treat the ailments as well as to address the sense of touch especially those with deteriorating senses. Its main focus will be to stimulate the sense of touch. It is apparent that the products designed can serve everyday functions so that the elderly can use them as much as possible without having to actually 'be in' therapy; they simply have to go through their daily routine and still be treated (Fig. 4).

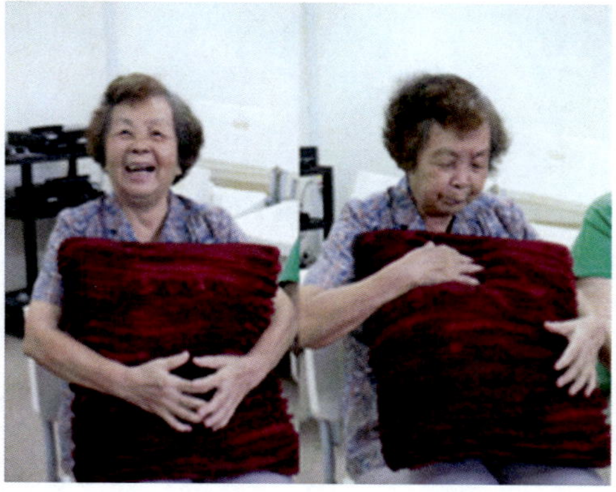

Fig. 5 Heated cushion

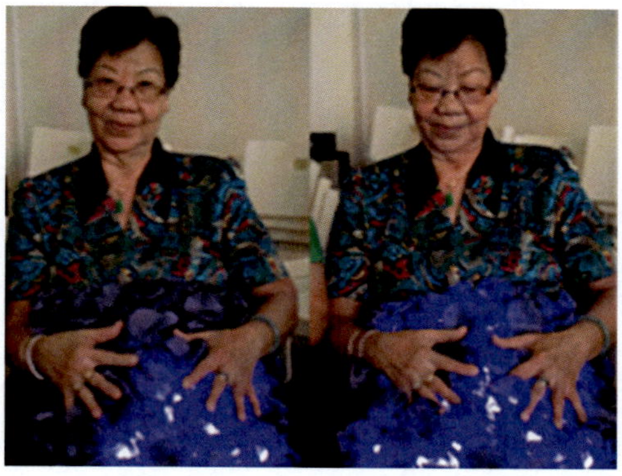

Fig. 6 LED light cushion

Conclusion

Textiles now have a new potential beyond fashion and aesthetics. For this research, it is found that textiles can be used as a source of alternate therapy for the geriatric patients and the elderly. This range of products could be added to the existing market of products that targets mainly on the tactile therapy for the target market.

Further tests will be carried out to assess the successfulness of the use of Tactile Textiles with embedded technology. These developments will be based on the field tests conducted with the aid of a geriatric health-care worker who can track the effectiveness of tactile therapy on the daily lives of the geriatric patients and the elderly.

We often disregard the possible benefits of tactile therapy, like how we disregard the needs of the elderly when it comes to designing for them due to the lack of understanding about the effects of aging. This project tries to merge both of those things.

'SMART textile' may very well be a new start for tactile therapy to infiltrate everyday lives, not only for the aged. It could also be beneficial for selected groups of users who need tactile sensory inputs in order for them to improve and enhance their daily lives.

In summary, tactile therapy through the use of three-dimensional structural textiles will provide a variety of tactile sensations and explorations using touch. The implementation of SMART embedded technology into the products of three-dimensional structural textiles will therefore generate more interest from the users. This will better equip the users with the appropriate therapeutic tactile sensory benefits and hence enhance holistic approaches that improve both the well-being and quality of their lives (Figs. 5 and 6).

Acknowledgement The author wishes to acknowledge the funding support for this project from Nanyang Technological University under the Tier 1 funding, Singapore Institute of Technology for the funding support of this conference and, last but not least, all the lovely aunties who have tested out the prototypes and Mr. Choo Jin Kiat, executive director of O'Joy Eldercare Centre.

References

1. V.L. Schmall, Sensory changes in later life (1991). Retrieved 22 Aug 2013, from http://ir.library.oregonstate.edu/xmlui/bitstream/handle/1957/15925/PNW196-fromArchive.pdf?sequence=1
2. Design Goteborg, Technology as design material. Retrieved 22 Aug 2013, from https://dru.tii.se/reform/projects/itextile/pillow.html
3. M. Mcinaney, Pet-therapy project launched in geriatric psych unit (1997). Retrieved 22 Aug 2013, from http://news.stanford.edu/news/1997/august27/paws.html
4. S. Braddock, M. O'Mahony, *Techno Textiles – Revolutionary Fabrics for Fashion and Design* (Thames and Hudson, London, 1998)
5. S. Braddock, M. O'Mahony, *Techno Textiles 2 – Revolutionary Fabrics for Fashion and Design* (Thames and Hudson, London, 2005)
6. G.K. Stylios, Editorial: SMART textiles special issue. Trans. Inst. Meas. Control, 213–214 (2007). Academic Search Premier, EBSCOhost (Accessed October 26, 2013)
7. L.V. Langenhove, C. Hertleer, Smart clothing: a new life. Int. J. Cloth. Sci. Technol. **16**(½), 63–72 (2004). Business Source Premier, EBSCOhost (Accessed October 27, 2013)
8. SEFAR PowerHeat, Retrieved 22 Aug 2013, from http://www.sefar.com/htm/609/en/SEFAR-PowerHeat.htm?Slider=Product+Features&Folder=1484711&pc3Scroll=0x499

Thermal Energy Storage of Polyester Fabric Coated with Paraffin Liquid as Microencapsulated Phase Change Material (PCM)

A.B.M. Dom, A.F. Mohd, N. Tulos, E. Nasir, and W.Y.W. Ahmad

Abstract

This research work involves the production of microencapsulated phase change material (PCM) in which paraffin liquid was used as the core component with two different concentrations of sebacyol chloride(SC) of 0.17 mol (Sample A) and 0.18 mol (Sample B) and hexamathylene diamine (HMD) (0.2) to make polyamide shell components. The microencapsulated PCM was characterized using Fourier Transform Infrared (FTIR) and Scanning Electron Microscopy (SEM). The thermal energy storage capacity was measured by Differential Scanning Calorimetry (DSC). The FTIR indicated that the microencapsulation process occurs due to the existence of alkyl group (C–H) and carbonyl group (C=O) in the spectra. DSC analysis shows that the paraffin started to melt at 20–30 °C with thermal energy storage capacity 0.06342 and 0.07925 J/g for Sample A and Sample B respectively. The results obtained show that the concentration of shell component affects the thermal energy storage of paraffin liquid.

Keywords

Microencapsulated PCM • Paraffin liquid • Thermal energy storage • Fourier transform infrared • Differential scanning calorimetry

Introduction

Energy storage depends on heat capacities that can be stored inside the material. The higher the storage capacities, the better it is. Thermal energy storage in phase change materials (PCM) offers the advantage of storing a large amount of energy in a small mass/volume. The energy density of the PCM can be as much as 50 times greater than heat of water. Added to this is that the stored energy is delivered over a narrow range of temperature [1]. Salaun [2] reported the use of melamine formaldehyde as shell for microencapsulated paraffin and found that the storage capacity was kept around 163–170 J/g, even after 13 thermal cycles.

In addition, PCM have long been used for thermal energy storage due to the large amount of heat absorption or release while undergoing phase change process with only small temperature variations [3]. This process occurs when the PCM reached its melting temperature during the heating process [4, 5]. Two most common groups of PCM are inorganic and organic compound. Inorganic PCM include salt hydrates, salts, metals, and alloys, whereas organic PCM are comprised of paraffin, fatty acids, or esters. Most organic PCM are noncorrosive and chemically stable and have little or no subcooling [3, 6].

Among many available organic PCM, paraffin is the most popular due to its special characteristics such as nontoxic, chemically inert, low cost ,and high storage heat capacity [7]. Instead of many promising characteristics on thermal energy storage, paraffin also shows some undesirable properties such as low thermal conductivity, noncompatibility with the plastic container, and moderate flammability [6].

A.B.M. Dom • A.F. Mohd • N. Tulos • E. Nasir • W.Y.W. Ahmad (✉)
Faculty of Applied Sciences, Universiti Teknologi MARA, 40450 Shah Alam, Selangor Darul Ehsan, Malaysia
e-mail: wanyunus@salam.uitm.edu.my

M.R. Ahmad and M.F. Yahya (eds.), *Proceedings of the International Colloquium in Textile Engineering, Fashion, Apparel and Design 2014 (ICTEFAD 2014)*, DOI 10.1007/978-981-287-011-7_9,

Many methods have been proposed for encapsulations of PCM such as in situ polymerization, complex coacervation, interfacial polymerization, and emulsion polymerization [7]. By using natural and biodegradable polymers, gum arabic–gelatin mixture, Onder et al. [8] have done encapsulated PCMs through complex coacervation. They reported that the heat storage performance of coacervates was very satisfactory for both 60 and 80 % dispersion ratios of PCMs.

Hence, this project was conducted with the aim to study the effect of encapsulated paraffin liquid coated on polyester fabric, especially in terms of heat storage capacities.

Materials and Methods

Materials

The project used Paraffin liquid as the PCM. Both hexamathylene diamine (HMD) and sebacoyl chloride (SC) were used in making polyamide as the shell for PCM. Polyvinyl alcohol (PVA) was used as a binder during the coating process and 100 % polyester fabric with the weight 1.41 g/cm^2, 3.5 mm in thickness, and density 9,100/in.^{2} with number of warp and weft was 130 and 70, respectively.

Methods

The microencapsulated PCM was produced using emulsion polymerisation technique in a 500 ml three neck round-bottomed flask equipped with a mechanical stirrer. Paraffin liquid and water was placed inside the flask. Next, 0.17 mol of SC (Sample A) was inserted slowly with 0.2 mol of HMD used in making polyamide as shell for the microencapsulated PCM. The reaction takes place for 1 h at temperature between 60 and 70 °C under stirring rate of 350 rpm. After the encapsulation process, the microcapsules of PCM obtained were filtered and washed with distilled water. Lastly, the PCM was dried in an oven at 40 °C for 24 h. The experiment was repeated using different concentrations of SC which was 0.18 mol (Sample B).

Characterizations of Microencapsulated PCM

Fourier Transform Infrared (FTIR)

Samples of the particles for infrared spectroscopy were grinded and mixed with Potassium Bromide (KBr) to produce pellets. Perkin Elmer Fourier Transform Infrared (FTIR) spectra in the transmission mode were connected to a PC and the results were recorded in which the number of scan was 16 and the resolution was 16 cm^{-1}.

Coating

Microcapsules were applied into 100 % polyester fabric by coating technique. The coating formulation was prepared by mixing homogenously microencapsulated PCM and binder PVA. The fabric was coated using Rollmac laboratory coating machine. Then, the coated fabric was dried at 100 °C for 5 min and cured at 150 °C for 3 min.

Scanning Electron Microscopy (SEM)

Scanning Electron Microscopy (SEM) was used to capture the image of microcapsules after coating on polyester fabric. The range magnification was at 500×, 1,000× and 5,000×.

Determination of Thermal Properties of Microencapsulated PCM

Differential Scanning Calorimetry (DSC)

NetzSch Differential Scanning Calorimetry (DSC) was used to determine thermal energy storage capacities which range from 0 to 100 °C and rate of heating 10 °C/min.

Result and Discussion

Fourier Transform InfraRed (FTIR)

The chemical compatibility among the components of the microencapsulated PCM was characterized by evaluating specific interactions between carbonyl compound from amide group and the PCM itself which is alkyl group from paraffin using FTIR spectroscopy technique. Figure 1 shows FTIR spectra of two different samples encapsulation of PCM. There is no significant difference between all the samples. C–H stretching peaks of alkyl group were found near 3,030–2,855 cm^{-1} and 1,485–1,415 cm^{-1}. These strong absorption bands of C–H were found in all the samples.

As for sample A and B, the FTIR spectra show that carbonyl compound (C=O) stretching were found near peaks 1,680 cm^{-1} and 1,665 cm^{-1}. It shows that strong bands for the C=O group appear in the microencapsulated PCM. These results indicate that there is a formation of microencapsulated PCM, with the existence of alkyl group (C–H) from paraffin and carbonyl group (C=O) from amide.

Scanning Electron Microscopy (SEM)

Figure 2 shows SEM images of microencapsulated PCM obtained after coating on polyester fabric. It shows that the tiny microencapsulated PCM from Sample A and Sample B were successfully coated on the surface of the polyester fabric. Microcapsules penetrated and attached into gaps between the fibers. This results may be due to friction that occurred during the coating process.

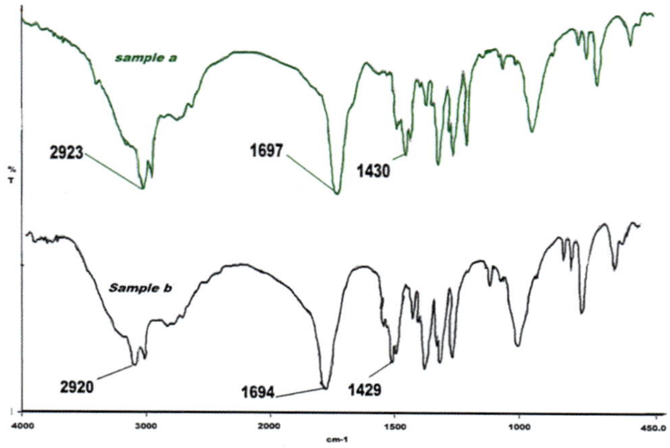

Fig. 1 FT-IR spectra of microencapsulated PCM for samples a, b, and c

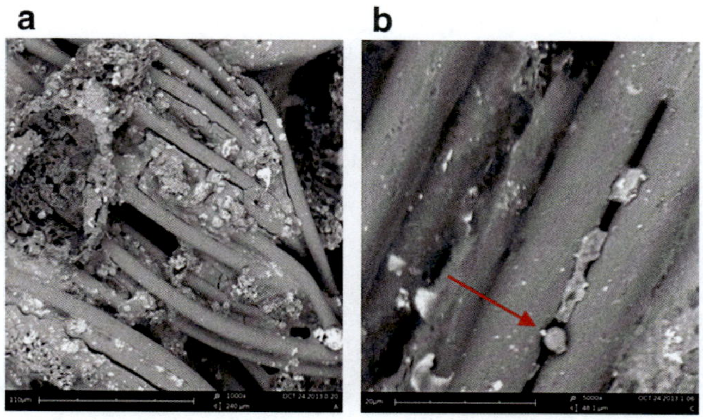

Fig. 2 (a, b) SEM image of microencapsulated PCM on polyester fabric

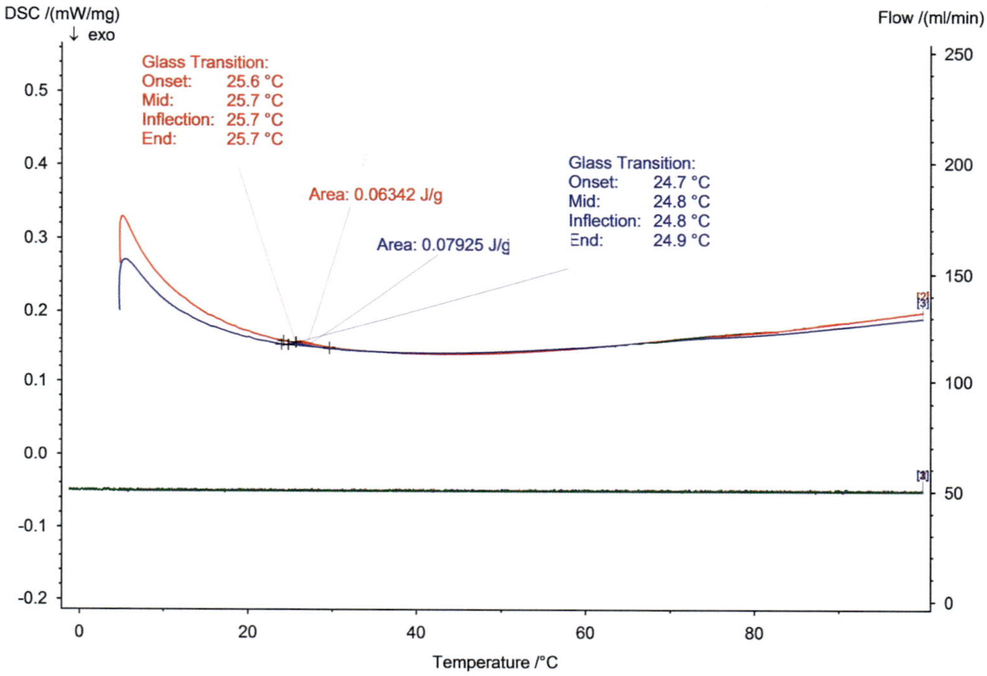

Fig. 3 DSC curve of microencapsulated PCM for samples a and b

Differential Scanning Calorimetry (DSC)

Figure 3 shows DSC curve of two samples of the microencapsulated PCM after coating on polyester fabric. Thermal properties obtained from the curve indicate that the microencapsulated PCM consist of paraffin liquid and started to melt at 20–30 °C. It also shows that the Sample A gave 0.06342 J/g less than Sample B with energy storage 0.07925 J/g. Different concentrations in making shell components give an effect on energy storage capacity of paraffin liquid, where 0.18 mol of SC (Sample B) give better result in thermal energy storage in comparison with 0.17 mol of SC (Sample A). Although the concentrations difference in making shell components was small, it does give an effect to thermal energy storage of paraffin liquid.

Conclusion

Microencapsulated PCM containing paraffin liquid was successfully prepared using emulsion polymerization process with two different concentrations for making shell of 0.17 and 0.18 mol of SC while 0.2 mol of HMD. Based on the FTIR results, the formation of microcapsules occurs due to the existence of carbonyl group (C = O) from amide and alkyl group (C–H) from paraffin. It also shows that the encapsulated PCM was successfully coated on the surface of the polyester fabric identified by desktop SEM. As for the thermal energy storage, DSC results indicated that the microencapsulated PCM consist paraffin with 0.18 mol of SC with 0.2 mol of HMD with Sample B giving better thermal energy storage in comparison with Sample A. It shows that the concentration of shell components also affect thermal energy storage of paraffin liquid.

Acknowledgment The authors would like to acknowledge Ministry of Education, Malaysia for the Fundamental Research Grant Scheme (FRGS), Research Management Institute (RMI), and Faculty of Applied Sciences, UiTM Shah Alam.

References

1. A.A. Sayigh, A. Hasan, Some fatty acids as phase change thermal energy storage materials. Renew. Energy **4**(I), 69–76 (1994)
2. F. Salaün, E. Devaux, S. Bourbigot, P. Rumeau, Development of phase change material in clothing part 1: formulation of microencapsulated phase change. Text. Res. J. **80**, 195–205 (2010)
3. G.H. Zhang, C.Y. Zhao, Thermal and rheological properties of microencapsulated phase change materials. Renew. Energy **36**(11), 2959–2966 (2011)
4. K. Koo, J. Choe, Y. Park, The application of PCMMcs and SiC by commercially direct dual-complex coating on textile polymer. Appl. Surf. Sci. **255**(20), 8313–8318 (2009)
5. G. Nelson, Application of microencapsulation in textiles. Int. J. Pharm. **242**(1–2), 55–62 (2002)
6. T. Kousksou, A. Jamil, T. El Rhafiki, Y. Zeraouli, Solar energy materials & solar cells paraffin wax mixtures as phase change materials. Sol. Energy Mater. Sol. Cells **94**(12), 2158–2165 (2010)
7. Z. Jin, Y. Wang, J. Liu, Z. Yang, Synthesis and properties of paraffin capsules as phase change materials. Polymer (Guildf) **49**(12), 2903–2910 (2008)
8. E. Onder, N. Sarier, E. Cimen, Encapsulation of phase change materials by complex coacervation to improve thermal performances of woven fabrics. Thermochem. Acta **467**, 63–72 (2008)

Fabric Mechanical Properties: Human Versus Machine Interpretation

S.A. Ghani, M.F. Yahya, and S.N. Dahalan

Abstract

Clothing comfort based on fabric mechanical properties was studied by making a comparison between subjective and objective evaluation. The subjective evaluation was done by conducting surveys to 100 working female respondents. The objective evaluation was done by testing the fabric mechanical properties which included drapability, sweat absorbency, stiffness and air permeability. The study aimed to find a correlation on the relationship between human's comfort interpretation based on hand touch and laboratory testing using mechanical action related to the fabric properties. It was found that good correlations of R^2 values were obtained between human and machine interpretation for stiffness, sweat absorbency and air permeability.

Keywords

Clothing • Comfort • Subjective evaluation • Objective evaluation • Drapability • Sweat absorbency • Stiffness • Air permeability

Introduction

Comfort is one of the important characteristics needed by consumers when buying cloths. The feeling of comfort in clothing is related between physical, psychological and sensorial perception [1]. Fabric comfort can be classified into three groups which include psychological, tactile and thermal [2]. Psychological comfort is not related with the fabric properties but more on the garment style, proper fit, fashion and suitability for the occasion. A tactile factor is related to the skin in contact with the fabrics or how the wearer feels when the skin touches the fabric [3]. Factors attributing to tactile property include types of fibre, yarn structure, fabric type, dimension characteristics, tensile properties, stiffness, elasticity and surface morphology or surface roughness. Fabric finishing also contributes to the comfort aspect especially in the usage of chemicals and types of finishing treatment [4].

Thermal property is determined by the movement of air, heat and moisture in and out of the fabrics [4]. Thermal property is influenced by the type of fibre, yarn, fabric and finishes [5]. Fabrics play an important role in maintaining the heat balance as it modifies the heat loss from the skin and at the same time altering the moisture loss from the skin [6].

The comfort property of fabrics has been the subject of interest for many researchers [7–9]. Unfortunately, limited studies were done on human's perspective towards fabric comfort. The human perspective is very important in decision-making during purchasing of their outfit [10].

Because of that, the main aim of the study is to investigate whether the human perspective is related and comparative or significantly different with the result of laboratory testing. The fabric comfort properties that were investigated include drapability, sweat absorbency, stiffness and air permeability.

S.A. Ghani (✉) • M.F. Yahya • S.N. Dahalan
Faculty of Applied Sciences, Universiti Teknologi MARA, 40450 Shah Alam, Selangor Darul Ehsan, Malaysia
e-mail: suzai710@salam.uitm.edu.my

M.R. Ahmad and M.F. Yahya (eds.), *Proceedings of the International Colloquium in Textile Engineering, Fashion, Apparel and Design 2014 (ICTEFAD 2014)*, DOI 10.1007/978-981-287-011-7_10,

Materials and Methods

Fabrics

Five woven fabrics with plain weave structures of rayon, cotton, polyester (plain and satin weave) and polyester/cotton fabrics were used. The fabrics were of light weight to medium weight in the range of 80–180 gm^{-2}. The fabrics were divided into two parts: one part for subjective evaluation or survey and the other part for objective evaluation or laboratory testing. All the samples used were white in colour since the purpose of this study was to know the human's perception on the fabrics based on texture and fabric feel which should not be influenced by colour psychology.

Objective Evaluation

Drapability

The Cusick drape tester was used in order to determine the drape coefficient of the fabric. The fabric was cut into circular samples using a 30 cm diameter template cutter. Each sample was placed in the centre pin of the drapability tester disk. Once the light on the tester was switched on, the shadow of the fabric drape appeared on the drape paper. The resultant shadow was traced and the shape of the shadow on the drape paper was cut. The cut-out of the draping paper was weighted and the value was divided with the uncut draping paper to get the drape coefficient. All the steps were repeated three times for the front and back surfaces for each sample and the average reading was calculated.

Sweat Absorbency

The sweat absorbency test was done by determining the time a drop of sweat flattened onto the surface of the fabric. Alkaline sweat was prepared by mixing 0.1 g of L-histidine monohydrochloride mono-hydrate, 1 g sodium chloride, 1 g crystallized disodium hydrogen orthophosphate, 200 ml distilled water and 0.1 N of sodium hydroxide together. A sample of fabric was placed on top of the beaker without any support at the centre of the sample. Then a drop of sweat was dropped in a distance of 1 cm from the top of the fabric. The time for the sweat to be fully flattened was taken using a stop watch.

Stiffness

All the samples were prepared for warps and weft direction with three replicates. The samples were cut according to the test template. A sample was then placed on the sliding board with the edge of the sample at the edge of the dotted point. The template was placed on the top of the sample and slides until the edge of the sample touched the black line. The readings of the edge touching the black line were taken four times and the average was calculated. Both edges of each sample were tested.

Air Permeability

A sample was firmly clamped in the rig of the selected test head or area. The machine was operated through a control panel that allows the operators to set the parameters needed. 20 cm^2 of testing area was chosen because it is the recommended testing area for woven fabric. The machine determined the resistance of the sample to the passage of airflow under constant preset air pressure.

Subjective Evaluation

A survey was conducted to 100 office working female respondents with the range age between 20 and 45 years old. The questionnaires and a sample book were given to the respondents during the survey. The questions were related with the fabric properties which were tested using appropriate testing equipments.

Samples of Questions

Arrange the fabric (in the sample book) based on the following fabric performance:

1. Excellent, 2. Good, 3. Moderate, 4. Fair and 5. Bad

Questions

1. That drapes well on your body
2. That absorbs your sweat quickly
3. That ease your body movement
4. That you feel suitable to wear in hot weather

Objective Versus Subjective Evaluation

Question 1 in the survey was to reflect the drape coefficient of the fabric using the Cusick drape tester. Question 2 was to determine the rate of sweat absorbency. Question 3 was regarding stiffness property of the fabric and Question 4 was to evaluate the air permeability.

The ranking for both objective and subjective evaluation was done in order to get R^2 values on the correlation between these selected properties.

Results and Discussion

All fabrics were tested for their basic properties such as weight, thickness and density as shown in Table 1. The weight of the fabric was in the range of between 80 and 140 gm^{-2} with the thickness in the range of 0.2–0.6 mm.

All the fabric mechanical properties which include drape coefficient sweat absorbency, for both alkaline and acidic,

Table 1 Fabric basic properties

Fabric ID	Fibre (weave)	Weight (gm^{-2})	Thickness (mm)	Density (EPI/PPI)
A	Rayon (plain)	105.33	0.282	110/55
B	Cotton (plain)	106.00	0.366	80/66
C	Polyester (plain)	140.33	0.569	148/76
D	Polyester (satin)	81.33	0.273	221/107
E	Poly/cotton (plain)	111.67	0.308	149/76

Table 2 Fabric comfort properties

Fabric ID	Drape coefficient (%)	Sweat absorbency (second)	Stiffness (mm)	Air permeability ($l\ min^{-1}$)
A	39.29	1.13	2.03	152.95
B	65.32	0.59	1.64	116.78
C	33.07	1.42	1.18	48.23
D	33.61	6.91	1.60	39.15
E	66.22	99.2	2.03	34.34

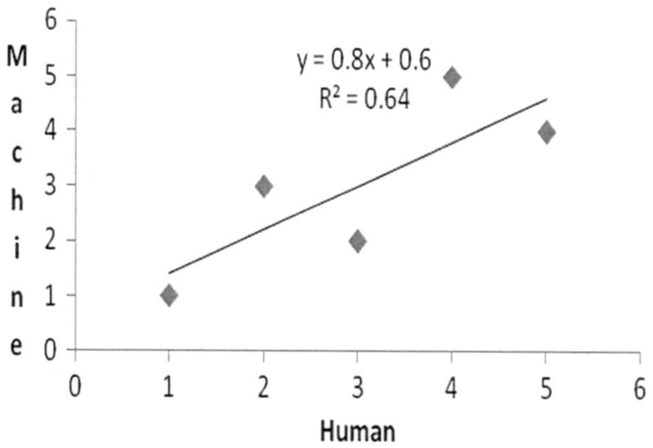

Fig. 1 Drapability

bending length for warp and weft and air permeability are shown in Table 2.

All the fabric comfort properties were considered objective evaluation and were compared to subjective evaluation in terms of their ranking. The ranking was based on the best properties, labelled as 5 and the lowest, labelled as 1. For drape coefficient, the lowest value was the best, the ranking also applied to sweat absorbency and stiffness. For air permeability, the highest value gave an indicator that air can pass through more efficiently. The highest reading was ranked as 5. All the ranking value for both subjective and objective evaluation was compared and shown in Figs. 1, 2 and 3.

R^2 for sweat absorbency, stiffness and air permeability was very good at the value of 0.81. The R^2 value of 0.81 showed that humans can do a good prediction on the comfort

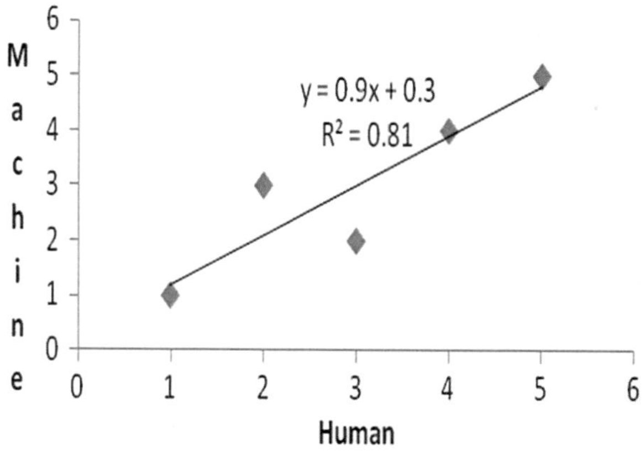

Fig. 2 Sweat absorbency

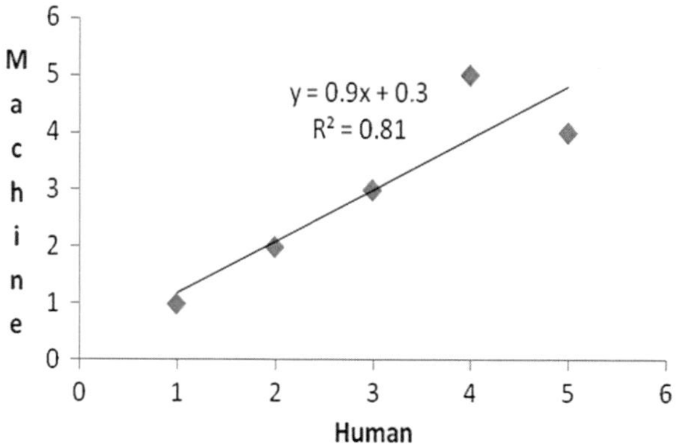

Fig. 3 Air permeability and stiffness

properties on sweat absorbency, bending property and air permeability just by looking and feeling the fabric by hand. The accuracy of human ranking on the five selected fabrics was close to the ranking that was done by using human interpretation. The drapability R^2 was 0.64, which is statistically significant but not as strong as the other three properties.

Conclusion

From the study, it can be concluded that humans can predict fabric properties such as drapability, sweat absorbency, stiffness and air permeability based on the hand touch and eye coordination. The survey was done on working females and contributed to a very good correlation with the results from the machine. The respondents were people experienced in buying and selecting fabrics especially for fabric sold in rolls. The study can be continued by conducting surveys to different sets of target groups and other fabric comfort properties such as water vapour and thermal properties.

Acknowledgement The authors gratefully acknowledge the Ministry of Higher Education (MOHE) for the Fundamental Research Grant Scheme (FRGS) and the Research Management Institute (RMI), Universiti Teknologi MARA (UiTM), for the management of the fund.

References

1. G.K. Tyagi, G. Krishna, S. Bhattacharya, P. Kumar, Comfort aspect of finished polyester-cotton and polyester-viscose ring and MJS yarn fabrics. Ind. J. Fibre Text. Res. **34**, 137–143 (2009)
2. A. Das, S.M. Ishtaque, Comfort characteristics of fabrics containing twistless and hollow fibrous assemblies in weft. JTATM **3**(4), 1–7 (2004)
3. R.K. Nayak, S.K. Punj, K.N. Chatterjee, Comfort properties of suiting fabrics. Ind. J. Fibre Text. Res. **34**, 122–128 (2008)
4. A.H. Taieb, S. Msahli, F. Sakli, A new approach for optimizing mechanical clothing tactile comfort. J. Adv. Res. Mech. Eng. **1-2010**(1), 43–51 (2010)
5. A. Mitra, A. Majumdar, P.K. Majumdar, D. Banerjee, Comparative analysis of regression and ANN models for predictind drape coefficient of handloom fabrics. Ind. J. Fibre Text. Res. **37**, 313–320 (2011)
6. M. Dhinakaran, S. Sundaresan, B.S. Dasaradan, Comfort properties of apparel. Ind. J. (Department of Textile Technology, Kumaraguru College of Technology, Coimbatore)
7. K.V.P. Singh, A. Chartterjee, Study on physiological comfort of fabric made of structural modified friction spun yarns: part II- liquid transmission. Ind. J. Fibre Text. Res. **35**, 134–138 (2009)
8. S.A.R. Zeinab, M. Saad, M. El Shinkery, I. Hanafy, Textile fabrics as thermal insulators. Autex Res. J. **6**(3), 148 (2006)
9. I. Frydrych, G. Dziworska, J. Bilska, Comparative analysis of the thermal insulation properties of fabrics made of natural and man-made cellulose fibres. *Fibres and Textile in Eastern Europe*, 2002
10. M. Mushtaq, *Clothing Comfort a Combination of Objective and Subjective Evaluation* (Technical University Liberce, Liberce, 2010)

Upper Fitness Personal Assistant: Body-Guts

C.W. Tan, S.W. Chin, A.W.H. Teo, W.X. Lim, and D.W. Goh

Abstract

This paper proposes Body-Guts, a dry-fit apparel that plays the role of monitoring the wearer's health and gives the user a reason to exercise more often. Nowadays, exercise is so essential that many do not pay much attention to it; since then, many people suffer from being overweight and from diseases that are caused by it. According to the World Health Organization (WHO), in the year 2010, 60 % of Malaysians with age above 18 are overweight which also causes the problem of heart attack, stroke, and diabetes. The Body-Guts is knitted with an electrocardiography (ECG) heart rate sensor and electromyography (EMG) muscle sensor. These sensors provide relevant data which will be analyzed in the Arduino microcontroller and then transmit significant information to a smartphone application via Bluetooth. Electrode pads will collect EMG signals which allow the monitoring of muscle stress via a graphical user interface (GUI) in mobile application, whereas a heart pulse sensor is used to obtain heartbeat rate for the calculation of calories burned. Furthermore, the mobile application also includes workout tutorial and a list of required nutrients together with a workout monitoring system based on EMG and calories burned. Throughout the ECG experiment, the ECG sensor results collected via mobile "Bluetooth spp" apps as compared to radial pulse have an accuracy of up to 99.83 %, whereas the EMG that uses a conductive fabric as adhesive gel achieves a waveform that demonstrates the forearm strength produced.

Keywords

Body-Guts • Upper Fitness • Sensors • Health • Body-Fit Apparel

Introduction

In the current generation we live in, the majority live a sedentary lifestyle, or better known as a couch potato. Living without physical exercise can be life-threatening by predisposing the victim to cardiovascular diseases which can be very common these days. Even so, people who exercise in an extreme manner can also experience cardiac arrest. Such a case happened recently in India where a 38-year-old TV actor Abir Goswami faced sudden death while working out on the treadmill [1]. According to cardiologist Dr. James O'Keefe [2], physical activity can be treated similarly as a medical treatment, which means it can be harmful if it is overdone. As a wrap-up, each individual needs to be well aware of the level of workout; as for the workout heart rate, sports people need to be familiar with their own heart limitation during workout as well.

To counter such problems, sports and medical monitoring products such as BioHarness 3 from Zephyr [3], fabric chest strap from NuMetrex [4], MyoLink muscle sensors from Somaxis [5], and many more had stood up to confront the existing issue. The NuMetrex heart rate monitoring system is built in with a heart rate sensor that is knitted into the apparel

C.W. Tan (✉) • S.W. Chin • A.W.H. Teo • W.X. Lim • D.W. Goh
School of Engineering, Science and Technology, KDU College Penang Campus, Penang, Malaysia
e-mail: cw_93@hotmail.com

M.R. Ahmad and M.F. Yahya (eds.), *Proceedings of the International Colloquium in Textile Engineering, Fashion, Apparel and Design 2014 (ICTEFAD 2014)*, DOI 10.1007/978-981-287-011-7_11,

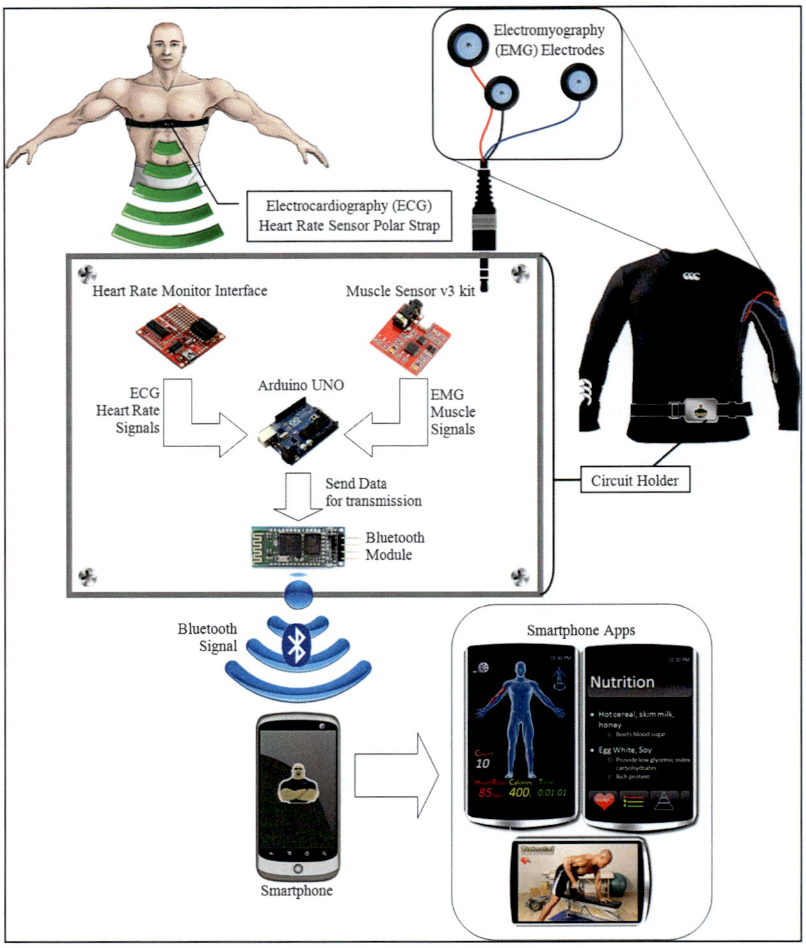

Fig. 1 Body-Guts block diagram

where an elastic chest strap is replaced with the apparel. The system requires a transmitter to transmit electrocardiography (ECG) signals to devices designed to display information such as calories burned based on the user's heart rate [4]. With only the calories burned data, the information it can deliver to the user is limited, and hence, information can be further improved by providing a nutrition list based on the calories burned data. Alternatively, the Somaxis MyoLink muscle sensor which is still under beta test uses an electromyography (EMG) electrode that is built in with a transmitter, which allows posture monitoring and prevention of repetitive strain injury via the user's mobile apps [5]. Instead of adhesive gel electrodes with the purpose of remaining in contact with the skin, the Somaxis MyoLink consists in its outstanding performance in terms of wireless muscle sensing but with cons of using nonreusable EMG electrodes.

In order to confront the aforementioned product feature limitation, the Body-Guts is proposed with the combination of EMG and ECG sensors. The EMG electrode of the Body-Guts is designed using conductive fabric as a replacement of the adhesive gel to achieve its permanent and water-proof characteristics. The ECG polar strap heart rate sensor is knitted in the dry-fit suit in use with its elasticity to keep the sensor in full contact with the skin. The received heart rate data is then transmitted to the Body-Guts apps and where calories burned is displayed and processed to provide suitable nutrient information to the users. The Body-Guts apps fitness system provides workout tutorials that are designed differently in 10 stages. It also monitors the user's heart rate condition during workout and triggers a warning when the user is near to critical condition. Lastly, the design comes with a wearable belt circuit holder which can be carried easily during workout.

Upper Fitness Personal Assistant

In order to pursue the Body-Guts goal, sensors have been reviewed and priorities have been made. These priorities towards the sensors are low power consumption, water resistive at its best, least bulky, and simple interfacing. The EMG v3 sensors [6] have been examined together with the polar strap heart rate sensor and conductive fabric to achieve the Body-Guts simple appearance as well.

Figure 1 shows the overview of the proposed idea, Body-Guts. The Body-Guts consists of a heart rate monitoring

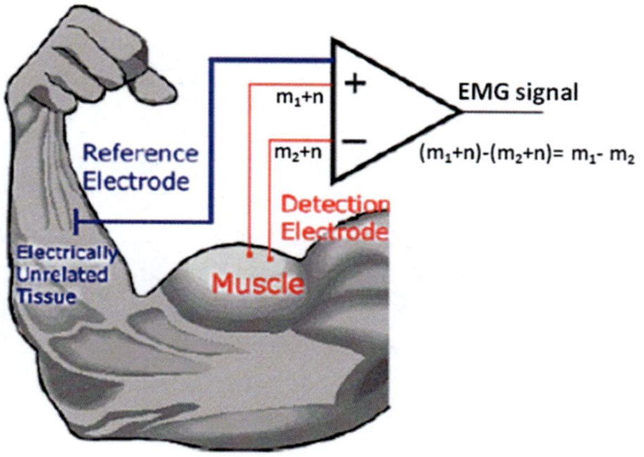

Fig. 2 Illustration of electrode location

system which works with a polar strap heart rate sensor that transmits ECG signal wirelessly to a heart rate monitor interface (HRMI) circuit that collects the heart rate data to the main microcontroller, Arduino UNO. The Body-Guts not only has the heart rate sensor; it also includes a muscle sensor which collects EMG [7] signals directly from the body in contact with EMG electrodes. These signals will be amplified and filtered in a muscle sensor v3 circuit board. The EMG signals will then be sent to the Arduino UNO. Signals that are collected from the Arduino UNO will then be transmitted through a Bluetooth module, which will be connected and received by the user's smartphone apps as shown in Fig. 1. The smartphone apps are developed using Android SDK which is involved in the workout tutorial, workout counts via EMG signals, and calories burned via the heartbeat rate. The calories burned information will then be processed to suggest nutrients suitable for the user.

Surface Electromyography Sensor Electrodes (sEMG)

The EMG sensor includes three electrodes which receive raw EMG signals to a third-party V3 muscle sensor circuit that amplifies, smoothens, and rectifies the signal for simpler further application. Each surface electrode consists of its own location on the human muscle. These locations determine the muscle difference regarding the differential amplifier of the V3 circuit. The diagram shown in Fig. 2 specified the placement of each location on the bicep. Basically, there are three electrodes with three different colors: red (+), blue (−), and black (ground reference). The red electrodes are placed on the muscle body, in this case, on the bicep, whereas the blue electrode is attached to the lower end of the muscle body that represents the negative part of the differential amplifier. The last electrode which is black colored will be placed on the nonmuscular part of the body, as referenced in Fig. 2, which is attached to the forearm. The reason why the red and blue electrodes are attached prior to the muscle body is because the muscle sensors determine the signals based on the difference between the most flexed muscle cells and the least muscle cell electrons produced by the muscle neurons. With these small electrical potentials, the amplifier is able to amplify and smoothen the signal of the muscle activity. As for the Body-Guts, the forearm muscle is prior since it detects the bicep muscle as well during flexion. The electrode placement is similar as described for the bicep, but with the ground, it is placed at the back of the forearm which produces the least electrical potential and can be isolated from the muscle flexion of the muscle body.

The equation $(m_1 + n)-(m_2 + n) = m_1-m_2$ shown in Fig. 1 determines the EMG signal collected from the differential amplifier where "m" is the EMG signal and "n" represents the noise signals [8]. The aforementioned V3 muscle sensor circuit is a differential amplifier that is designed with its user-friendly interface that is compatible with the Arduino UNO. Hence, the raw signals obtained from the human skin which consist of unpredictable noises are being filtered and, at the same time, amplified and rectified.

Conductive Fabric

The human skin surface consists of a layer of dry skin cells, which act as input impedance that may range up to several thousand ohms and affects the EMG signals obtained. In order to confront this problem, the adhesive gel is introduced where it allows better connection between the skin and the electrodes which provides better signal. Unfortunately, the adhesive gel does not reach the Body-Guts requirement that prior on water-resistive characteristic. Hence, the conductive fabric is chosen as a substitution for the adhesive gel. The chosen conductive fabric is made out of nickel, copper, and cobalt coated with nylon fabric which consists of high conductivity and high flexibility, good adhesive which is also washable. The electrical conductivity or surface resistivity of the conductive fabric is less than 0.1 Ω/sq [9].

Polar Electrocardiography Heart Rate Sensor (ECG)

Electrocardiography, also known as ECG, is generally known as the heartbeat signal which is obtained from the human body surface by attaching ECG electrodes. The heart rate is obtained by electrical potential difference generated

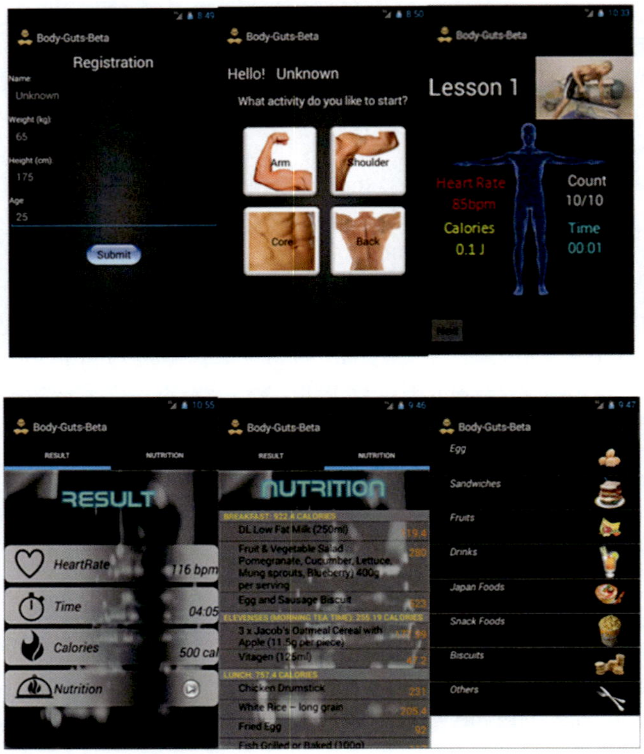

Fig. 3 Screenshot of smartphone application

from each individual cell from the heart and their sequence of activation. This means the subsequent altering heart posture within the body generates the electrical force that reflects the electrical activity of individual cells which represents the individual's heartbeat. This technology is implemented in Body-Guts due to its stability and the accuracy it provides compared to the photoplethysmogram (PPG) that estimates the skin blood flow using infrared light.

Body-Guts Software Development with Android SDK

Android developers offered an open source firmware, the Android Software Development Kit (Android SDK) [10], and it is compatible with Arduino Bluetooth function. The smartphone application is used to display the processed signal from Arduino and display it to the user. The apps will request for height, weight, and age as depicted in Fig. 3 in order to calculate calories and critical heart rate. Once the users key in the required information, the apps will turn to the exercise main page to allow the user to select the workout activity for different body parts (Fig. 3). In a workout, the user is guided through a series of three activity lessons. In every activity, a tutorial video will be provided and the user is required to meet the target count to proceed to next lesson by exercising based on the video. Bluetooth function in smartphone is enabled to receive signal from Arduino during the activity. EMG and ECG signal collected from Arduino will be used to determine the workout counts and also the calories burned through further calculation. The heart rate and calories are changing real time and displayed to the user. After the workout, different nutrition lists will be suggested to the user based on the heart rate and calories, along with food database as a diet reference.

Human Calories Burned Value and Nutrition Balance

During exercise, the human muscles must burn calories to fuel the contractions. The conversion of calories from the stored nutrient state to the form that can be burned by the muscle cells is achieved, during aerobic exercise, through the process of cellular respiration, which requires oxygen. The delivery of oxygen through the bloodstream to muscle cells is directly related to the heart rate. Therefore, through the heart rate-based calorie burn formula, the rate of burning calories during aerobics can be calculated based on the average heart rate while performing the exercise. Essentially, with the increased exercise intensity, the body muscles must burn more calories, and so the heart must beat faster to provide the oxygen necessary to convert those calories to the form of energy that can be burned by the muscles. However, nutrition balance through the calories consumed must equal the calories expended for a person to maintain the same body weight. Consuming more calories than expended will result in weight gain; meanwhile, consuming fewer calories than expended will result in weight loss [11].

Equation for the determination of calorie burn [12]:

$$\text{Male} = \Big(\big(-55.0969 + (0.6309 \times HR) + \big(0.1988 \times W\big) + (0.2017 \times A)\big) \div 4.184 \times 60 \times T\Big) \tag{1}$$

$$\text{Female} = \Big(\big(-20.4022 + (0.4472 \times HR) + \big(0.11263 \times W\big) + (0.074 \times A)\big) \div 4.184 \times 60 \times T\Big) \tag{2}$$

Where:
HR = Heart rate (in beats/minute)
W = Weight (in kilograms)
A = Age (in years)
T = Exercise duration time (in hours)

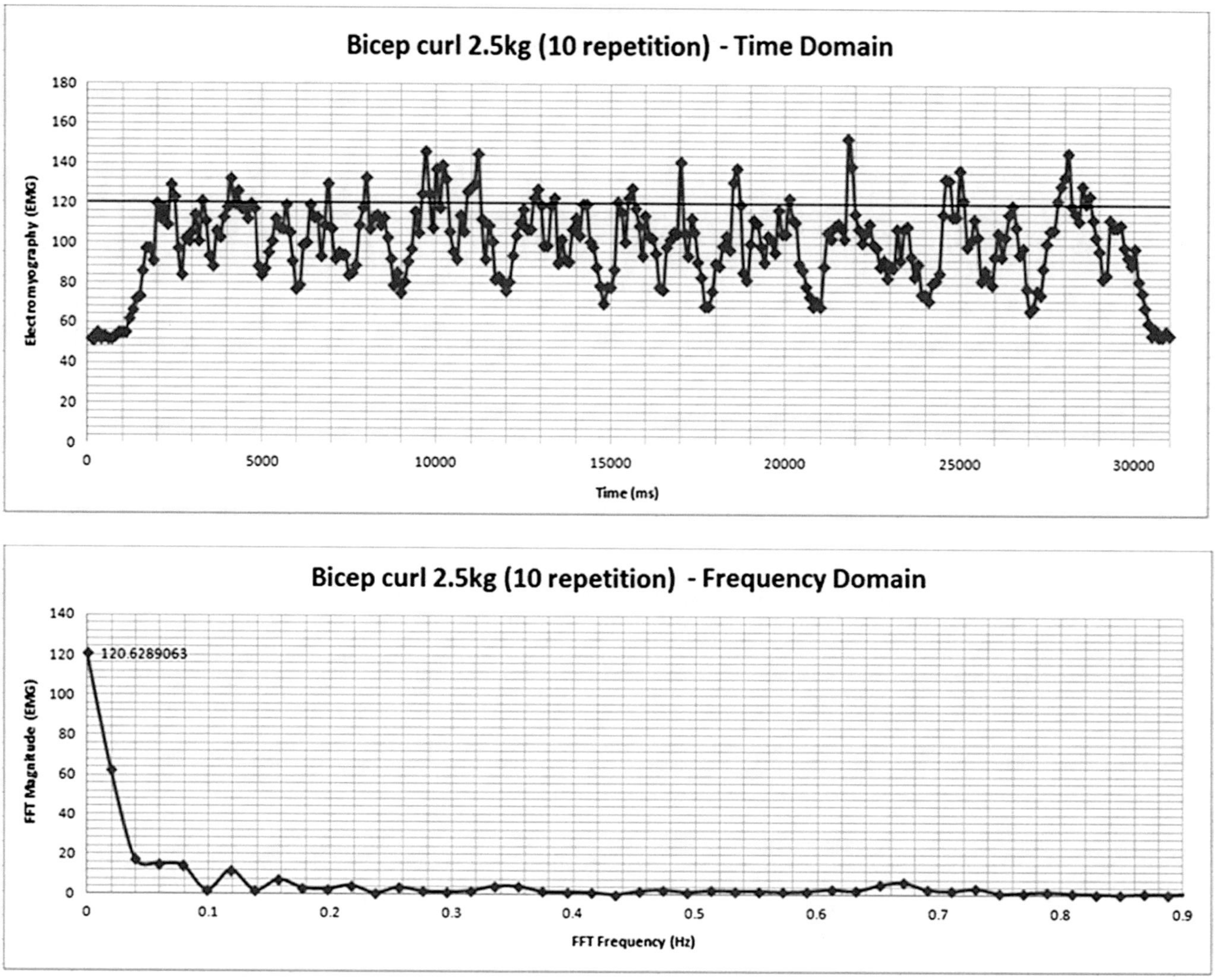

Fig. 4 Time and frequency domain graph when the dumbbell weighs 2.5 kg

Simulation Results and Analysis

Effectiveness Test of Muscle Sensor

The first experiment demonstrated the effectiveness of muscle sensor, by evaluating the EMG signal produced from the user. The experiment aim is to investigate the relationship between weight handle by user with the muscle signal produced. From the relationship, the threshold value for counting can be determined based on the different set of weights handled by the biceps. Before conducting the experiment, the EMG sensor electrodes are attached to the user bicep. During the experiment, the user is requested to work out with a dumbbell with different sets of weight of 2.5, 5, and 7.5 kg. The circuit power supply is remained on throughout the experiment. Meanwhile, the muscle signal received in real time is converted to digital value, and the value is recorded and being tabulated. At post experiment, the time domain and frequency domain graph is plotted from the data collected as shown in Figs. 4, 5, and 6.

In the first round, a 2.5 kg weight dumbbell is used for bicep curls workout where its signal is filled with noises. The reason why the signal is unclear is because the weight has least effects on the muscle, which means the potential difference between the two forearm muscles are low and hard to identify; hence, the EMG responded at low alternative signals. Compared to the second round, a 5 kg dumbbell experiment resulted with a convincing EMG signal waveform. From the waveform result, a total of 10 pulsating signals can be observed where each pulsate waveform consists of two peaks; this occurs due to the ups and downs of bicep curls workout where lifting the weight forms a higher EMG and lowering the weight forms another smaller spike. As the

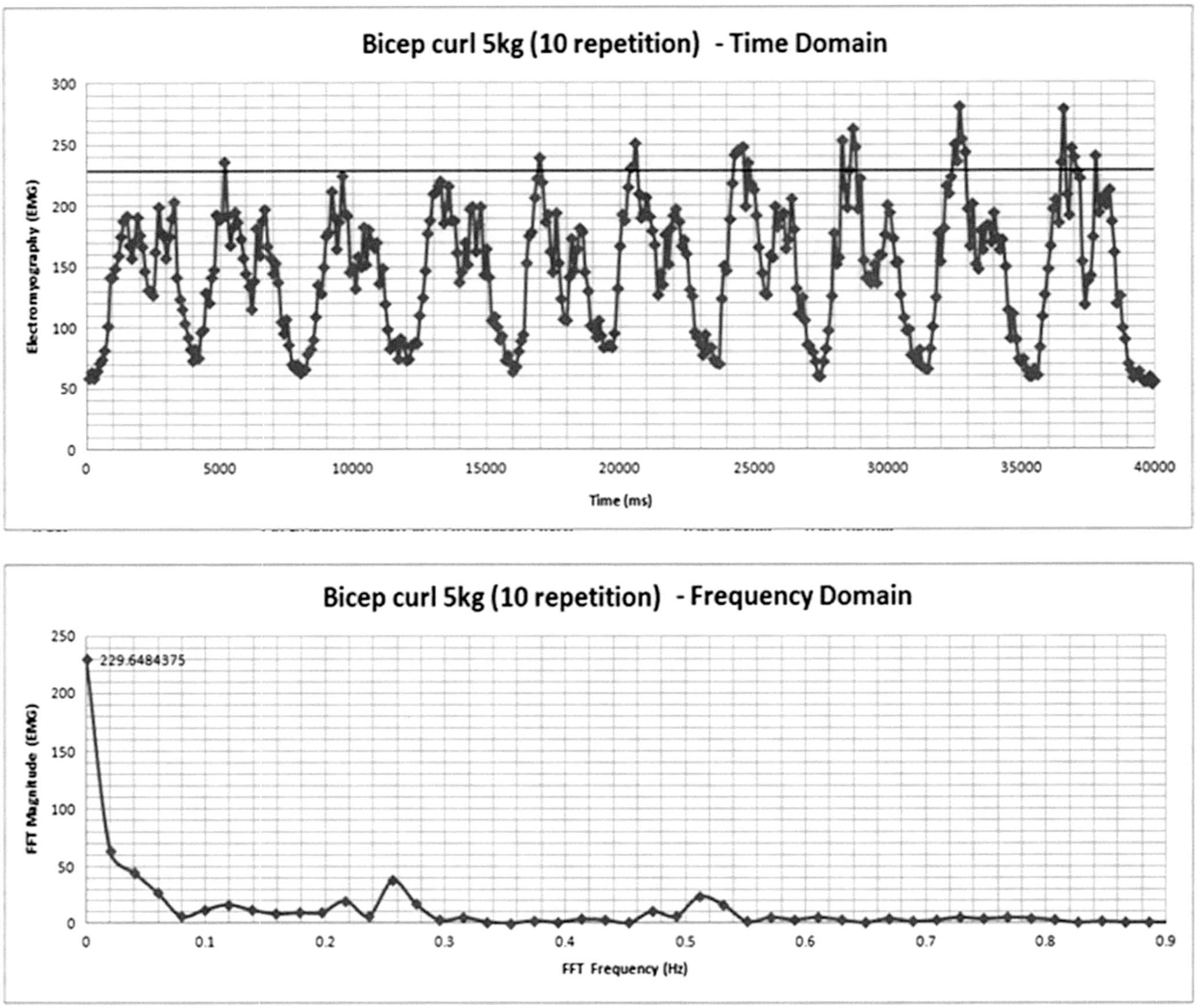

Fig. 5 Time and frequency domain graph when the dumbbell weighs 5 kg

workout counts increase, the EMG signal increases; this is because the muscle gets tired and the brachioradialis begins stressing, causing larger potential difference and thus larger EMG signals. The third experiment uses a 7.5 kg dumbbell with a result of higher EMG signals but a similar waveform pattern as compared to the weighted 5 kg dumbbell experiment. The EMG rises because larger strength is produced from the muscle while taking control of the weight.

As an overall wrap-up, the threshold EMG for 2.5 kg is 120, for 5 kg is 229, and for 7.5 kg is 324. The value is obtained from the frequency domain of the result. As the weight or difficulty of the workout increases, the strength required to accomplish the workout increases, thus producing higher potential difference which is the EMG signal.

Robust Test of Heart Rate Sensor

A heart rate sensor's accuracy is evaluated in the second experiment, by comparing the signal given by the sensor itself and manually counting the user's radial pulse. As similar as the equipment, the user remains at sitting rest position for a minute. The circuit power is switched on from the beginning to receive heart rate signal until the end of the experiment. The user is also requested to count their own radial pulse rate during the experiment. The Bluetooth module is designed to transmit each signal in 0.1 s, which will result in 600 ECG signals in 60 s. After the process of the experiment, the 600 values from the simulation result which are recorded from the "Bluetooth spp" apps are taken

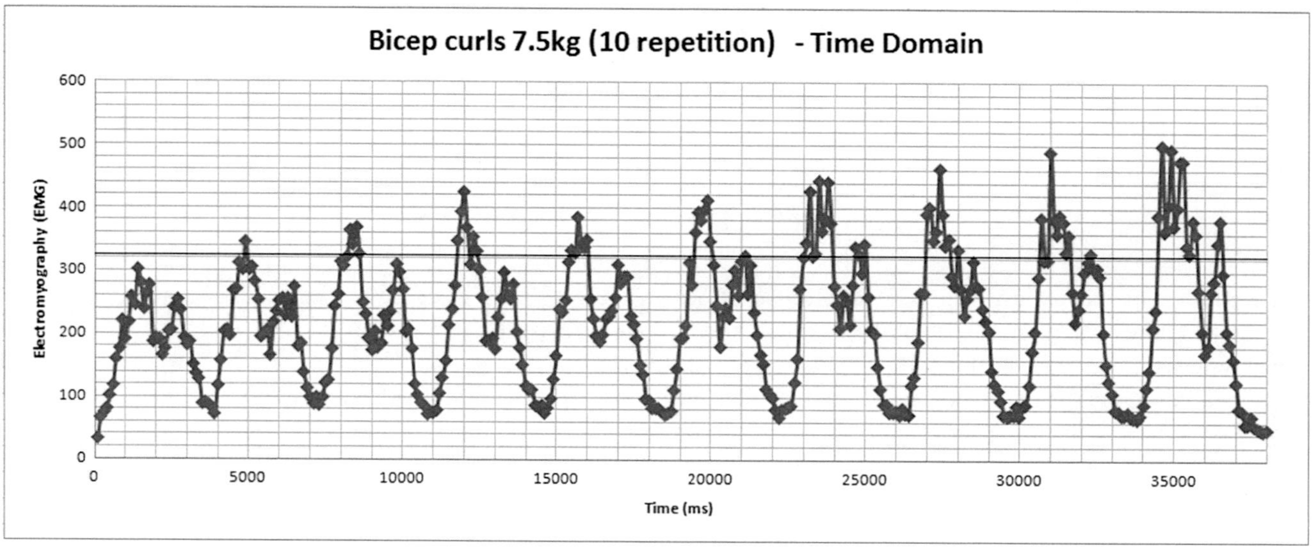

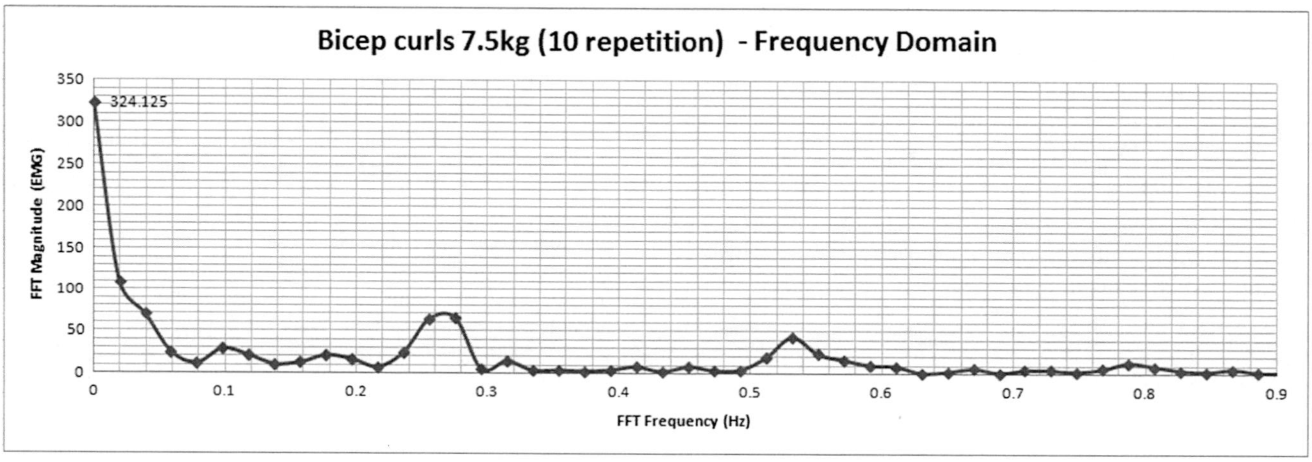

Fig. 6 Time and frequency domain graph when the dumbbell weighs 7.5 kg

Table 1 Measurement given by ECG sensor and radial pulse

Tests	ECG sensor (BPM)	Radial pulse (BPM)
1	61.59	61
2	68.96	69
3	71.41	72
4	72.94	73
5	67.91	68
6	68.94	68
7	70.22	71
8	67.89	68
9	70.76	71
10	66.21	67
Average	*68.68*	*68.8*

Total average ratio of ECG and radial pulse = 68.68:68.8

Table 2 Heart rate data during the push-up activity

	Heartbeat rate (BPM)	
Push-up (counts)	Before activity	After activity
10	71	113
20	79	119
30	81	133
35	83	142

to calculate the average ECG value. As a result, the average is compared with the user's final count, and then the accuracy is calculated. The experiment procedure is repeated for ten rounds, and the results are recorded as shown in Table 1.

$$\text{Accuracy}\,(\%), \frac{68.68}{68.8} \times 100 = 99.83\,\% \tag{3}$$

As for the radial pulse measurement, the pulse is measured at the same time the ECG collects its signals within the 60 s. The final heartbeat counts is then recorded. Lastly, the accuracy of the ECG sensor with comparison to radial pulse is calculated with a promising value of 99.83 %.

Table 3 Heart rate data during the stairs running activity

	Heartbeat rate (BPM)	
Stairs stairs	Before activity	After activity
Ground to 1st floor	73	125
Ground to 2nd floor	75	141
Ground to 3rd floor	79	150
Ground to 4th floor	84	163

The third experiment involved two physical activities with the aim to test the performance of the heart rate monitoring system during workout. First, the user is required to record the initial heart rate before the activity. When the activity is carried on, the results are recorded via "Bluetooth spp" apps with Bluetooth connection to the device. Final heart rate is recorded once the activity is complete. The sequence of activity below is repeated once the user is rested and ready to start (Tables 2 and 3).

For the push-up test, the heart rate increases as the push-up counts increase. Similar to the stairs running activity, the heart rate gradually rises as the level of workout increases. In a common sense of looking at the situation, the heart rate rises as body movement increases. This is because active muscle cells require more oxygen and energy delivered from blood which is pumped from the heart; hence, the larger the muscle intensity, the higher the oxygen and energy it requires, which leads to faster heartbeat [13].

Conclusion

With the completion of Body-Guts, experiments have been conducted to ensure its stability and the robustness of its performance. Every signal received from the sensors is crucial towards identifying the calories burned to monitoring the workout posture; hence, ECG sensors from the sensors to the smartphone apps has its accuracy up to 99.83 %, and the EMG sensor using conductive fabric is able to identify the strength a person can hold up to during workout. Intense activity such as stairs running from the experiment of ECG has taken into experiment to test the stability of the ECG sensor and also the accuracy of heart rate at different conditions. As a conclusion, the accuracy of the signals received is at its promising level where an individual can identify and monitor their workout data via smartphones.

References

1. K. Behrawala, How much is too much in the gym?, in *The Times of India* (2013)
2. J. O'Keefe, (2010) [cited 23 Nov 2013] Available from http://www.cardiowellnesscenter.org/Providers/James_OKeefe_MD.aspx
3. BioHarness 3 (2012) [cited 2013] Available from http://www.zephyranywhere.com/products/bioharness-3/
4. Wearable Technology. The NuMetrex heart rate monitoring system (2013) [cited 23, 2013] Available from http://shop.numetrex.com/wearable-technology/
5. MyoLink muscle sensors (2012) Available from http://somaxis.com/2012/07/20/myolink-muscle-sensors-2/
6. Muscle sensor v3 (2013) [cited 2013] Available from http://dlnmh9ip6v2uc.cloudfront.net/datasheets/Sensors/Biometric/Muscle%20Sensor%20v3%20Users%20Manual(1).pdf
7. J.S Huff, Electromyography (EMG). Introduction [cited 22, 2013] Available from http://www.emedicinehealth.com/electromyography_emg/article_em.htm
8. C.J.D. Luca, in *Surface Electromyography: Detection and Recording*. Electrode and Amplifier Design (Delsys, 2002)
9. Ni/Cu/Co fabric tape (2013) Available from http://www.lessemf.com/fabric.html
10. Welcome to the android open source project (2013) Available from http://source.android.com/
11. Balancing calories (2011) 31 Oct 2011 [cited 23 Nov 2013] Available from http://www.cdc.gov/healthyweight/calories/
12. Heart rate based calorie burn calculator (2013) [cited 23 Nov 2013] Available from http://www.shapesense.com/fitness-exercise/calculators/heart-rate-based-calorie-burn-calculator.aspx
13. Exercise intensity: why it matters, how it's measured (2013) 5 Mar 2011 [cited 23 Nov 2013] Available from http://www.mayoclinic.com/health/exerciseintensity/SM00113/NSECTIONGROUP=2

Surface Appearance Changes of Bio-finished Knitted Fabric

E. Nasir, M.S.R.M. Khair, N. Tulos, A. Musa, A. Baharudin, and S.A. Ghani

Abstract

Pilling is one of the common problems that affect the surface appearance change of fabrics. In this study, the application of chitosan as a finishing agent was conducted using pad-dry-cure method on acrylic 1 × 1 rib and cotton interlock fabric. The main aim of this study was to determine the surface properties of knitted fabrics, before and after the chitosan-based treatment. The physical characteristics of the treated fabric such as the percentage of weight loss in abrasion, pilling resistance and its microscopic view were investigated. The pilling resistance was observed using pilling standard grade after some revolution of pilling box. The abrasion resistance was analysed by the percentage of weight loss of the fabric after rubbing. Generally, it can be concluded that the application of chitosan improved the pilling resistance of 1 × 1 rib acrylic as compared to interlock cotton fabric. The pilling formation on the treated fabric's surface was lower in comparison with the untreated fabric.

Keywords

Chitosan • Knitted fabric • Pilling resistance • Abrasion resistance • Textile surface change

Introduction

Knitted fabric has its own uniqueness that makes it popular in the garment or clothing industry which cannot be found in other form of 2-dimensional textile material. Knitted fabrics are popular as the technology has improved tremendously over the past decades with different kinds of knitted structures, modified yarns and design of knitting instruments [1]. However, knitted fabrics are always associated with problems such as dimensional stability and surface appearance changes [2].

In general, a knitted fabric has some comfort properties, more elastic than woven or nonwoven, and cling well to the body to inhibit movement. However, during its utilization, pills form on the surface of the fabric and remain on the surface of the end product which worsens its aesthetic value [3]. The pills become a main problem for acrylic knitted fabric, especially when friction forces result in fibre degradation which affects the appearance of the fabric [4]. The formation of pills, which is called as pilling, occurs when the fibres are pulled out from the fabric creating a fuzzy surface. Eventually, the entangled fibres will form pills, making the fabric look old and worn out. It is often difficult to restore a garment with fabric pilling to its original condition, especially wool knitwear [5, 6].

Knitted materials are prone to surface appearance change due to its loose structure as compared to woven fabrics. In the context of fabric construction, knitted materials are prone to pill more than woven due to the morphology of knitted structure which is composed of tiny bent yarns called loops. Knitted fabric is also loose in structure where the mobility of the loops has caused the fibre in the yarn being pulled out when in contact with moving rough surface and pressure.

The pilling tendency of knitted fabrics can be affected by the characteristics of the fibre in the yarn, spinning process, construction of the fabric as well as the finishing treatment applied before garment make-up process [7]. At the fibre

E. Nasir (✉) • M.S.R.M. Khair • N. Tulos • A. Musa • A. Baharudin • S.A. Ghani
Faculty of Applied Sciences, Universiti Teknologi MARA, 72000 Kuala Pilah, Negeri Sembilan, Malaysia
e-mail: eryna@ns.uitm.edu.my

M.R. Ahmad and M.F. Yahya (eds.), *Proceedings of the International Colloquium in Textile Engineering, Fashion, Apparel and Design 2014 (ICTEFAD 2014)*, DOI 10.1007/978-981-287-011-7_12,

level, the fibre strength, diameter, length and curvature have an impact on the rate of fuzz formation [3]. In terms of fibre composition, Busiliene, Lekeckas and Urbelis [6] wrote that fabric from polyester fibre has better pilling resistance than viscose, bamboo and cotton. It was also reported that cotton fabric has the roughest surface that increased after repeated launderings [8, 9]. In the current study, acrylic and cotton are selected due to large consumption of these fibres in clothing and apparel industry.

It was mentioned earlier that finishing treatment affects pilling behaviour of knitted materials. As the environmental awareness has increased, many researchers have worked on eco-friendly finish such as enzymatic treatment [10–12]. In this study, chitosan was used to treat the fabric through pad-dry-cure technique. Chitosan was selected as this material has given encouraging results in improving other properties of cotton such as antimicrobial and durable press as well as for surface modification treatment [13, 14]. The application of chitosan and its effect on acrylic and cotton fabric were examined to see if there was any possibility for the same material to give good pilling properties. Chitosan was chosen due to its non-toxic, biodegradable and biocompatible criteria, whereby this environmentally friendly method would add value to the final product.

Methods

Materials

Finished 1 × 1 rib fabric from 100 % acrylic and interlock fabric from 100 % cotton were used. The stitch density for 1 × 1 rib acrylic and interlock cotton was 22 cpi × 28 wpi and 43 cpi × 58 wpi, respectively.

The chitosan solution was prepared by stirring 10 g of chitosan flakes in 1,000 ml of aqueous acetic acid solution and left overnight at room temperature [14]. Then, the solution was filtered to remove any insoluble materials. 1 % concentration of chitosan was the source of the finishing agent that would be used in pad-dry-cure process.

Fabric Treatment

The fabrics were padded using pad mangle with wet pick-up of 80–85 %, 3 psi padding pressure and speed of 2 m/min. The fabrics then were dried at 80 °C for 5 min and cured at 150 °C for 3 min.

Table 1 Pilling standard rating

Standard	Description
Standard 5	No visual change
Standard 4	Slight surface fuzzing
Standard 3	The test sample may exhibit either or both of the following: (a) Moderate fuzzing (b) Isolated fully formed pills
Standard 2	Distinct fuzzing and/or pilling
Standard 1	Dense fuzzing and/or pilling which covers the sample

Fabric Laundering

Quickwash SDL ATLAS was used to wash the fabric samples, conducted according to AATCC Test Method 124 – 1996 [16]. The washings were repeated for 5, 10, 15 and 20 times. The goal of this procedure was to compare the degree of pilling, taking into consideration the mechanical action applied on the fabric surface during washing (not during the wearing stage). It was also done to investigate the performance of the chitosan coating on the surface change of the tested materials after several washings.

Fabric Tests

Fabric Pilling

The sample tubes were placed inside the pilling box for 5,000, 10,000, 15,000 and 20,000 revolutions. The samples were then evaluated by comparing the appearance with the pilling standards of 1–5. This method was conducted according to BS 5811 – 1986 [18]. Standard 5 gives the best pilling rating and Standard 1 gives the lowest rating as stated in Table 1.

Abrasion Resistance

The abrasion resistance test procedure was conducted according to BS 5690 – 1988 where the test samples were rubbed with abrasive materials under certain pressure [17]. The weight of the fabric sample was recorded before and after 20,000 rubbings.

Microscopic View

The fabric samples which underwent 20,000 revolutions of pilling box were viewed under a Leica microscope to observe the surface changes of the fabric.

Results and Discussion

Fabric Pilling

The results for this test are tabulated in Tables 2 and 3. Only the results of 20,000 revolutions and 10,000 revolutions for acrylic and cotton, respectively, were presented as they had the most changes after the treatment process.

It was found that acrylic fabric pilled the most at 20,000 revolutions of the pilling box. This was due to the nature of acrylic spun yarn which has great hairiness on its surface [15]. The friction between the acrylic surfaces has made the fibres to come out and made the fabric appear fuzzy. The structure of 1 × 1 rib also has contributed to the poor pilling as this structure is basically loose as compared to interlock.

However, the pilling resistance of acrylic was improved after the chitosan treatment. This was proven by the increased pilling grades as shown in Table 2. After the treatment, the acrylic fabric has moderate or slight surface fuzzing as compared to distinct pilling before the treatment. The application of chitosan on this fabric has removed the protruding fibres on the fabric and hence reduced the number of pilling formation on the fabric surface.

In comparison with acrylic, the chitosan treatment has given very minimum improvement to the interlock cotton fabric (Table 3). The pilling grades recorded for the treated test samples were 2 after 20 washings, a slight improvement than the untreated samples with dense pilling that covered the fabric surface (grade 1) after 15 washings.

A very little change made by the chitosan treatment was influenced by the fabric structure itself. A compact and tight interlock structure has caused the yarns to be more in contact with each other, with more fibres being raised on the fabric surface during washings. These fibres were eventually transformed into pills during the rotations of the pilling box.

Table 2 Pilling grade for acrylic after 20,000 revolutions

Fabric	0 washing	5 washings	10 washings	15 washings	20 washings
Untreated acrylic	3	2	2	2	2
Treated acrylic	4	4	3	3	3

Table 3 Pilling grade for cotton after 10,000 revolutions

Fabric	0 washing	5 washings	10 washings	15 washings	20 washings
Untreated cotton	2	2	2	1	1
Treated cotton	2	2	2	2	2

Abrasion Resistance

The overall results for the abrasion resistance are presented in Fig. 1. It was found that the percentage of weight loss for acrylic decreased after the application of chitosan treatment. The unwashed sample showed the greatest weight loss (9.44 %) due to the presence of protruding fibres on the yarn surface which were removed during rubbings. Washings have made the acrylic fibres further removed from the fabric surface due to the mechanical action during washing process, and hence the weight losses recorded for 10 washings and 20 washings were lower than the unwashed sample.

In comparison with the untreated acrylic, the weight loss for treated acrylic was smaller (3.04 %). Chitosan treatment has improved the fabric appearance which caused the fibres to retain on the surface. However, the fabric weight loss increased with the additional number of washings as the fibres were near to be completely removed from the fabric surface.

Although the pilling grades for cotton were poor for both untreated and treated samples, the percentage of fabric weight loss decreased, which represented a better abrasion resistance. The pills of fibre formed on interlock fabric were trapped in its compact structure without being removed during rubbings and washings which contributed to the decrease in fabric weight loss.

Microscopic View

Figure 2 shows the microscopic view for the untreated and treated acrylic after 20,000 revolutions of pilling box. The fibres on fabric surface have been entangled to form pills, and there was severe hairiness on the fabric protruded from the yarn. It can be seen in Fig. 2a where the fibres have filled up the stitch holes. In contrast, the treated fabric in Fig. 2b shows clearer stitch holes. The presence of chitosan in acrylic has improved the fabric appearance by preventing the formation of pills on the surface of the fabric.

Microscopic view of cotton fabric in Fig. 3 shows that the treated fabric (b) has less hairiness and clearer appearance. The stitch holes were less filled up with the protruding fibres as compared to the untreated sample (a). Visually, chitosan has made the yarn smoother and hence reduced the number of fibre filling up the stitch holes.

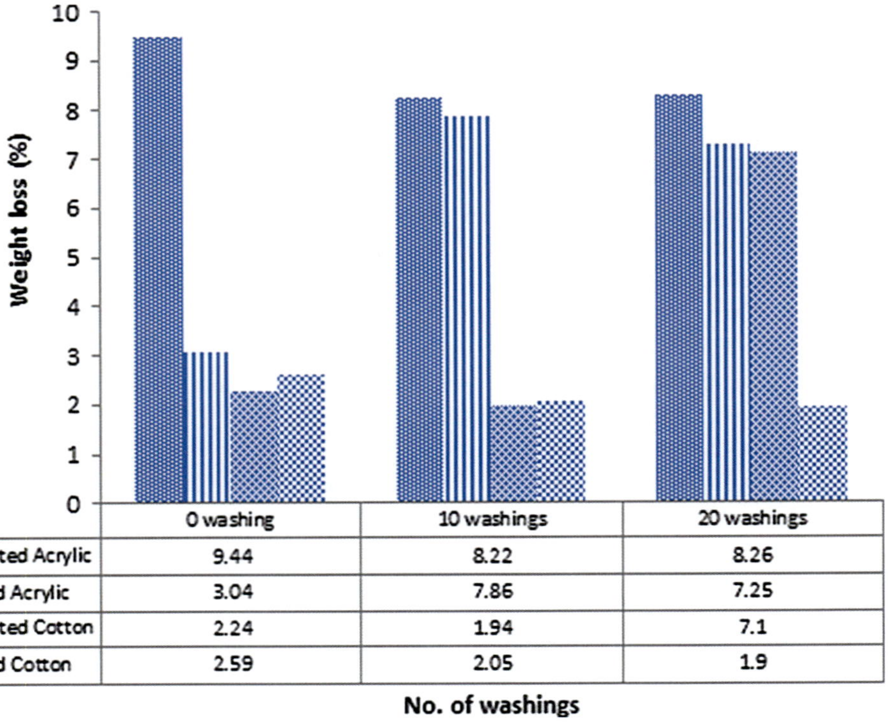

	0 washing	10 washings	20 washings
Untreated Acrylic	9.44	8.22	8.26
Treated Acrylic	3.04	7.86	7.25
Untreated Cotton	2.24	1.94	7.1
Treated Cotton	2.59	2.05	1.9

Fig. 1 Percentage of weight loss for acrylic and cotton after 20,000 rubbings (untreated and treated)

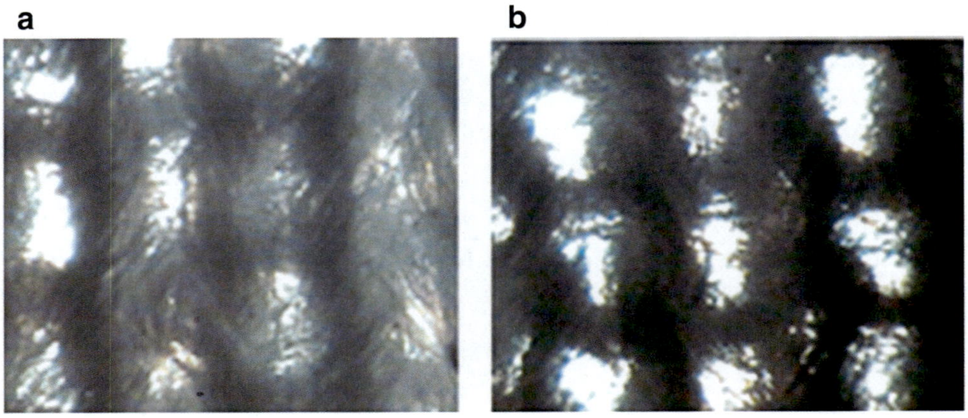

Fig. 2 Microscopic view of untreated (**a**) and treated acrylic (**b**)

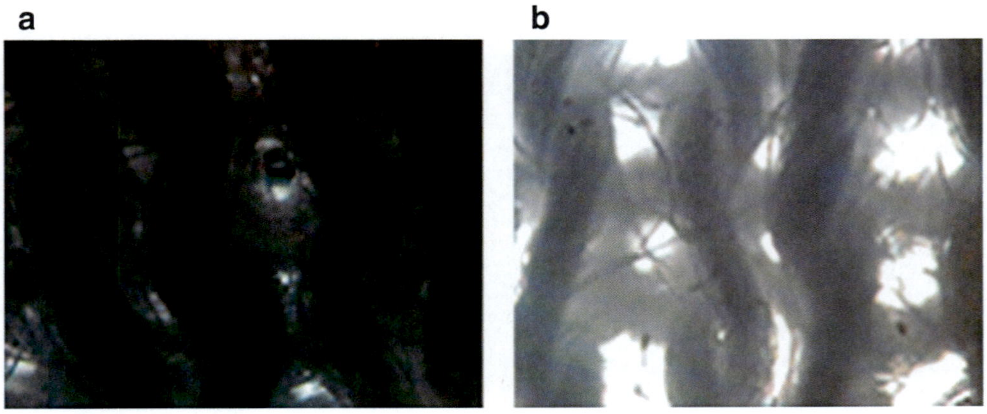

Fig. 3 Microscopic view of untreated (**a**) and treated cotton (**b**)

Conclusion

The results from this study showed that the application of chitosan using pad-dry-cure method was effective on acrylic knitted fabric. The formation of pills and hairiness on fabric surface has decreased for the treated acrylic, and the percentage of weight loss has also reduced. In contrast, the treated fabric just gave a slight improvement on the fabric's appearance properties. Chitosan has not been effective as a finishing agent with pad-dry-cure method on interlock cotton fabric as most of the treated fabrics showed very little enhancement on the appearance of fabric. Hence, there should be more improvement in terms of materials or application techniques in order to overcome pilling problems and finally enhancing the serviceability of the fabric especially with knitted structure.

Acknowledgement The authors gratefully acknowledge the Ministry of Higher Education (MOHE) for the Research Acculturation Grant Scheme (RAGS), UiTM, and the Research Management Institute (RMI) for the management of the fund.

References

1. M. Yanilmaz, B.F. Alev, F. Kalaoglu, A study on the influence of knit structure on comfort properties of acrylic knitted fabrics. Smartex Res. J. **1**(1), 88–92 (2012)
2. A.K.M. Mobarok Hossain, Prediction of dimension and performance of finished cotton knitted fabric from knitting variables. Fascicle Text. Leathe. **xiii**(2), 99–103 (2012)
3. G. Busiliene, K. Lekeckas, V. Urbelis, Pilling resistance of knitted fabrics. Mater. Sci. (Medziagotyra) **17**(3), 297–301 (2011)
4. Y. Can, Pilling performance and abrasion characteristics of plain-weave fabrics made from open-end and ring spun yarns. Fibres Text. East. Eur. **16**(1), 81–84 (2008)
5. N. Foster, What causes fabric pilling?, *WiseGeek*. Retrieved 10 Oct 2012, from www.wisegeek.com
6. M.Q. Tusief, N. Mahmood, M. Saleem, Effect of different anti pilling agents to reduce pilling on polyester/cotton fabric. J. Chem. Soc. Pak. **34**(1), 53–57 (2012)
7. I. Jazinska, Assessment of a fabric surface after the pilling process based on image analysis. Fibres Text. East. Eur. **17**(2), 55–58 (2009)
8. N.A. Kotb, Changes in knitted cotton/polyester fabric characteristics due to domestic laundering. J. Am. Sci. **8**(5), 677–682 (2012)
9. A. Hasani, Effect of different processing stages on mechanical and surface properties of cotton knitted fabrics. Ind. J. Fibre Text. Res. **35**, 139–144 (2010)
10. S. Jevsnik, D. Fakin, L. Heikinheimo, Z. Stjepanovic, Changes in a knitted fabric's surface properties due to enzyme treatments. Fibres Polym. **13**(5), 371–379 (2012)
11. G. Manonmani, V. Chettiar, Suitability of compact yarn for manufacturing of eco-friendly processed weft knitted fabrics. JTATM **6**(3), 1–18 (2012)
12. N.E. Özdil, N.E. Özdoğan, T. Öktem, Effects of enzymatic treatment on various spun yarn fabrics. Fibres Text. East. Eur. **11**(4), 58–61 (2003)
13. X.D. Liu, N. Nishi, S. Tokura, N. Sakairi, N. Chitosan, Coated cotton fiber: preparation and physical properties. Carbohydr. Polym. **44**, 233–238 (2001)
14. H.C. Yang, W.H. Wang, K.S. Huang, M.H. Hon, Preparation and application of nanochitosan to finishing treatment with anti-microbial and anti-shrinking properties. Carbohydr. Polym. **79**, 176–179 (2010)
15. S.L. Paek, Pilling, abrasion and tensile properties of fabrics from open-end and ring spun yarns. Text. Res. J. **59**(10), 577–583 (1989)
16. AATCC Test Method 124, *Appearance of Fabrics after Repeated Home Laundering* (Front Loading Washing Machine, 1996)
17. BS 5690, *Standard Test Method for Abrasion Resistance of Textile Fabrics* (Martindale Abrasion and Wear Tester, 1988)
18. BS 5811, *Standard Test Method for Pilling Resistance of Textile Fabrics* (ICI Pilling Box Tester, 1986)

The Performance of Tenun Pahang Using Various Weft Yarn

E.L.Z. Engku Mohd Suhaimi, J. Salleh, S.A.A. Ghani, M.F. Yahya, and M.R. Ahmad

Abstract

An investigation on the performance of Tenun Pahang fabric properties using new yarns was conducted. Five types of yarns was used as weft which were Tencel, Bamboo, Spun Silk, Polyester, and Polyester Coolmax. All fabrics were woven using silk warp yarns. The new weft yarns were evaluated against Tenun Pahang made from 100 % silk. Seven properties were evaluated including weight, thickness, thread density, abrasion resistance, crease recovery, drapability, and stiffness. The results show that the Polyester Coolmax yarn gives the best properties in term of its density, abrasion resistance, stiffness, and crease recovery angle.

Keywords

Tenun Pahang • Tencel • Spun silk • Polyester • Bamboo • Coolmax polyester • Fabric properties

Introduction

In Malaysia, Tenun Pahang seldom draws attraction compared with Songket and Batik fabrics. The young generation nowadays seems to be less interested in this kind of weaves. Normally, this art is appreciated by those who are really concerned in fashion [1].

Tenun Pahang is one of the famous legacies that belong to Malay ethnics and it is also well known due to its beautiful arts that have motif, subject, color, methods, and instruments. These features have made these arts to stand until today [1]. Unfortunately, as time goes by, these arts are forgotten and it is not possible that 1 day this art will be gone. Tenun Pahang needs to be exposed and commercialized so that it can turn out to be one of the heritages that represent the Pahang state [2].

Kain Tenun Pahang was believed to originate from Sulawesi in 1669 which the port in the place named Makassar was invaded by the Dutch. This caused a migration of the local Bugis far from the country. They finally landed in Pahang known then as Inderapura [3]. Kain Tenun Pahang was introduced by one of the chiefs from Bugis named Keraing Aji and held a respected title of Tok Tuan. He introduced a hand-woven fabric woven on the Malay frame loom to the local folk in Pahang [4].

The Pahang Development Authority or Perbadanan Kemajuan Negeri Pahang (PKNP) had taken the initiative to revive and promote Tenun Pahang since 1995 to keep this traditional heritage alive [5]. Tenun Pahang was conferred royal status "Diraja" by HH Tengku Abdullah Sultan Ahmad Shah, the crown prince of Pahang on the 8th of May 2006. In 2008, HH Tengku Puan Pahang, Tunku Hajah Azizah Iskandariah lauched a blueprint on Tenun Pahang which saw the establishment of an institute specially for Tenun Pahang managed by Perbadanan Kemajuan Kraftangan Malaysia [6]. Recently in 2012, PKNP established a Centre of Excellence for Tenun Pahang Diraja in Kampong Soi Kuantan, officiated by the Chief Minister of Pahang [7].

E.L.Z. Engku Mohd Suhaimi • J. Salleh (✉) • S.A.A. Ghani • M.F. Yahya • M.R. Ahmad
Faculty of Applied Sciences, Universiti Teknologi MARA, 40450 Shah Alam, Selangor, Malaysia
e-mail: jamilsal@salam.uitm.edu.my

M.R. Ahmad and M.F. Yahya (eds.), *Proceedings of the International Colloquium in Textile Engineering, Fashion, Apparel and Design 2014 (ICTEFAD 2014)*, DOI 10.1007/978-981-287-011-7_13,

Tenun Pahang have been woven using silk since the beginning until today but lately, Polyester is also used in Tenun Pahang especially for souvenir items. Silk is an expensive material and thus new material or alternative material is needed to lower down the cost without compromising quality.

New technology in textile has made available fibers with better properties which were available to be explored. Tencel fiber, for example, is the latest in rayon family; it has good dimensional stability and washing stability, low shrinkage in most cases with wash properties [8]. This fiber also has good drape properties and the section of the fiber is round making it luster, soft, and smooth. It is also moisture permeability, easy dyeing, bright color, and wearing likes as cotton.

Bamboo yarn, another newly developed fiber, has good durability, stability, tenacity, antibacterial and deodorizing in nature, incredibly hydroscopic which absorbs more water than other conventional fibers [9]. Another alternative yarn used in the project is Coolmax fiber which has good wrinkle and shrink resistant, hygroscopic, machine washable and dryable, stay soft and retain their shape, pulls moisture away from the skin [10].

Other alternative yarn used in the project is spun silk which is soft, but they are less lustrous than reeled silk, spun silk requires more twisting to hold the short fibers together [11]. Spun silk has less strength and elasticity than reeled silk; it represents all the general characteristics of reeled silk [12].

The fifth material used in the project is Polyester which is the most used synthetic fibers because Polyester fibres are versatile, easily modified, and suitable to be blend with most of the fibers. This fiber has an excellence abrasion resistance and strength, but the pilling of a Polyester fabric can cause a problem to the fabric. Polyester is also known for its low water absorbency [13].

There have been limited publications in the development of Tenun Pahang in terms of materials and techniques except for the work currently done with some modifications using handloom from Thailand and also introducing new designs such as "ikat". Therefore, this project investigates some of the properties of Tenun Pahang fabric using new yarns which are Tencel, Bamboo, Modal, and Coolmax Polyester in order to produce Tenun Pahang fabric with better performances compared with the present yarns. It is hoped to establish a benchmark in terms of standard specifications and performance for Tenun Pahang.

Experimental Method

Weaving of Tenun Pahang Using New Yarns

The Tenun Pahang fabric was produced at the Centre of Excellence for Tenun Pahang Diraja at Kampung Soi, Kuantan, Pahang which specializes in making Tenun Pahang fabric.

The Tenun Pahang fabrics were produced using alternative weft yarns which were Modal (second-generation rayon), Tencel (third-generation rayon), Bamboo yarn, and Coolmax yarn.

Figure 1 shows the flow chart of the processes involved in the project.

Testing of Physical and Mechanical Properties

There were eight (8) types of testing done on the Tenun Pahang fabrics, three on physical testing and five on mechanical testing:

Physical Testing

Weight

This test follows MS ISO 5084-2003 and ASTM D 3779-1996 standard. A standard cutter was used and the mass of the fabric was measured using a weighing balance and recorded as g/cm^2.

Thickness

This test follows ASTM D 3776-96/2002 standard. The thickness of the fabric was measured using a thickness gauge apparatus and recorded as mm.

Thread Density

This test was done in accordance to MS ISO 7211/2-2003. The density was counted by using counting glass equipment. It is the number of ends and picks per inch or per centimeter of a woven fabric. The counting was done with the help of pick counter. The results were recorded.

Mechanical Testing

Abrasion

The equipment used for abrasion testing was Martindale Abrasion Tester. This test follows the standard method of BS 5438-1988. The Tenun Pahang fabrics were cut into four

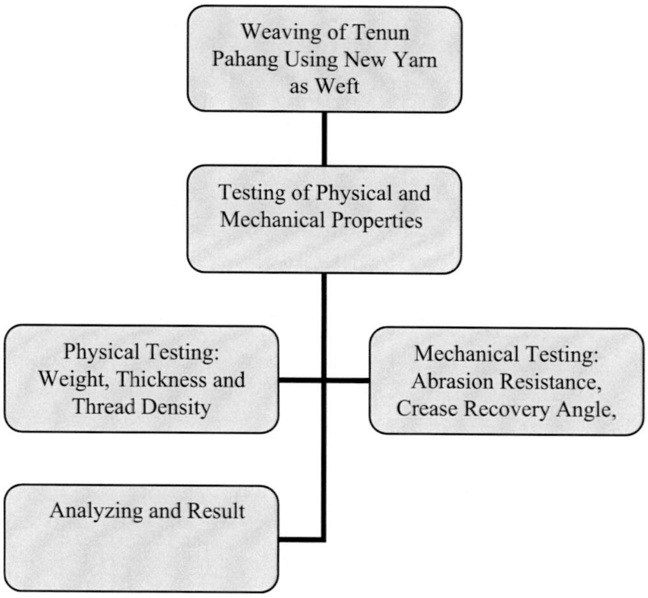

Fig. 1 The flow process of the project

samples per fabric with the diameter of 38 mm. A force of 9 kPa was used on the fabrics. The results were obtained by weighing the fabric sample before and after testing for 5,000 number of rubbings.

Pilling

The equipment used for Pilling testing was ICI Piling Boxes. This follows the standard of BS 5811-1986. Each of the Tenun Pahang fabrics was cut into the size of 125 mm × 125 mm.

Stiffness

The stiffness of the fabric was tested to determine the bending length and flexural rigidity of the fabric. This test was done following the standard method of ASTM D 1388–96/2002. The equipment used was the SDL Stiffness Tester.

Drapability

The purpose of the test is to determine the drape coefficient of fabric. This test was done in accordance to BS 5058–1997. The equipment used for the drapability testing was the Cusick Drape Tester.

Crease Recovery Angle

The purpose of this test is to determine the crease recovery angle of the fabric. The equipment used was the SDL Crease Recovery Angle Tester. The test follows the AATCC 66–2003 standard method. Ten (10) warp samples and ten (10) weft samples of 40 mm × 15 mm each were prepared. The test was repeated for the next 19 samples.

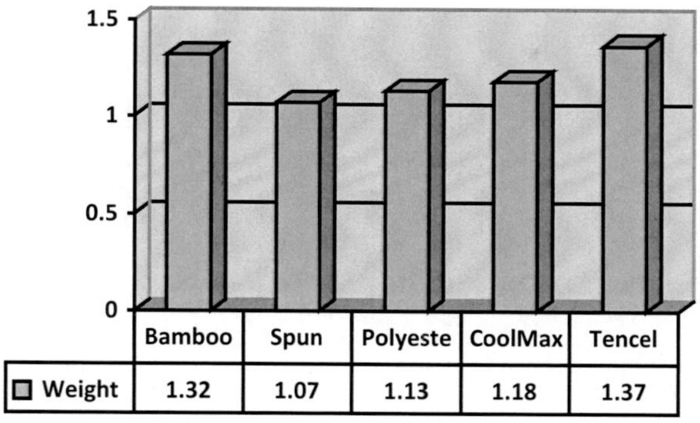

Fig. 2 Fabric weight result

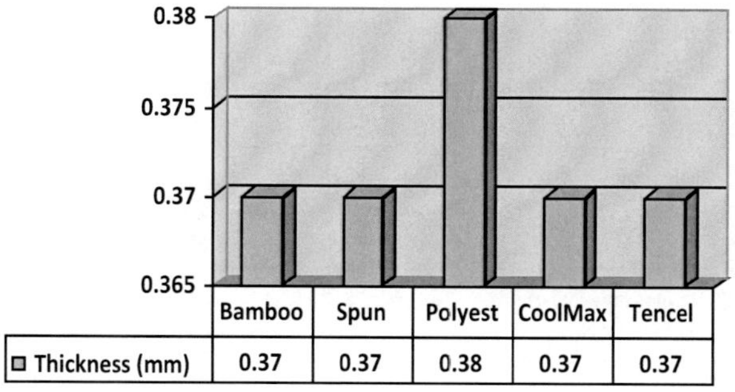

Fig. 3 Fabric thickness

Results and Discussion

Eight (8) testing were evaluated on the Tenun Pahang which were three (3) on physical testing (weight, thickness, and thread density) and another five (5) on mechanical testing (abrasion resistance, crease recovery, drapability, pilling, and stiffness).

Physical Testing

Fabric Weight

Figure 2 shows the results of fabric weight. The results show that Tencel fabric sample was the heaviest among the fabrics with 1.37 g/cm^2 followed by samples made from the Bamboo yarns, 1.32 g/cm^2. Polyester and Polyester Coolmax give a result of 1.13 and 1.18 g/cm^2, respectively, which makes Polyester to be lighter than Polyester Coolmax. The lightest fabric among the fabric sample was spun silk with 1.07 g/cm^2.

Thickness

Figure 3 shows the results of fabric thickness for each fabric with different type of weft yarn used. The result shows that all the fabrics have almost similar thickness. This was probably due to the effect of using the same warp yarn. However the

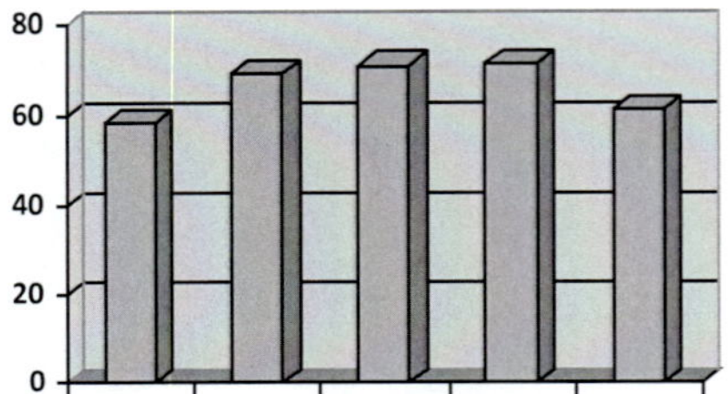

	Bamboo	Spun Silk	Polyester	CoolMax	Tencel
Density Weft (Ppi)	58.2	69	70.8	71.4	61

Fig. 4 Fabric thread density

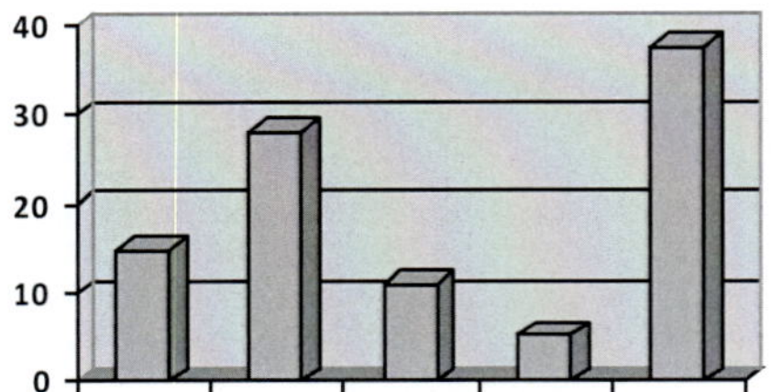

	Bamb	Spun	Polyes	CoolMax	Tencel
Percentage (%) of Weight Loss	14.9	28.2	10.9	5.2	37.5

Fig. 5 Percentage of weight loss

standard deviation of Bamboo and spun silk is a little high compared to others.

Thread Density

Figure 4 shows the results of fabric density for each fabric with different type of weft yarn used. From the figure, the density of the warp yarns varies with the sample from Bamboo yarns having the least picks per inch. The highest weft yarn thread density is the Polyester Coolmax yarn. This is because Polyester Coolmax is fine yarn which makes the fabric to be compact. This makes the fabric becomes smoother than the other fabrics. Samples of the fabric from Polyester yarn gave a thread density of 70.8 picks per inch which was almost similar to Polyester Coolmax.

Mechanical Testing

Abrasion Resistance

Figure 5 shows the result of abrasion and pilling for each fabric with different type of weft yarn used. From the result obtained, Polyester Coolmax shows the best result which is only 5.2 % of weight loss. This means that the fabric can withstand extreme rubbing with a rough surface.

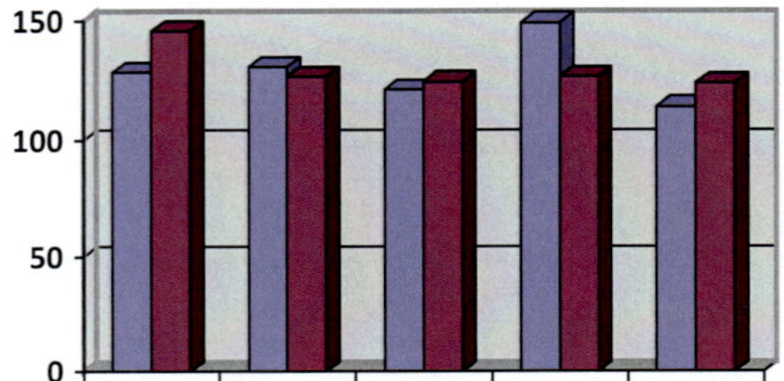

	Bamb	Spun	Polyes	CoolMax	Tence
F - to - F	128.6	130.8	121.2	149.2	114.2
B - to - B	145.8	126.6	124.6	126.6	124

Fig. 6 Fabric crease recovery angle

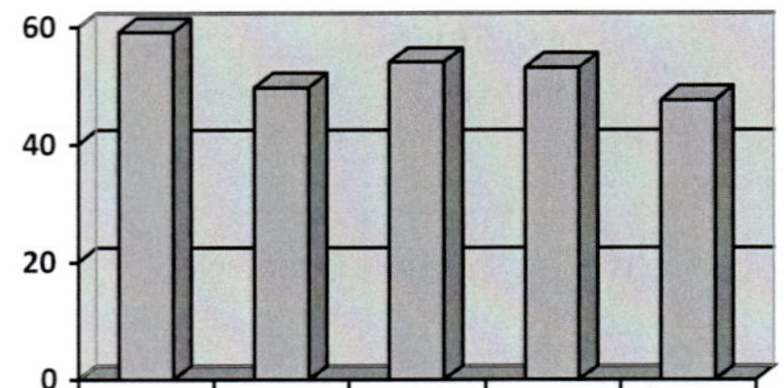

	Bambo	Spun	Polyes	CoolMax	Tencel
Drape Coefficient (%)	59.13	49.77	54.09	53.13	47.58

Fig. 7 Fabric drape coefficient

Crease Recovery Angle

Figure 6 shows the result of crease recovery angle for each fabric with different types of weft yarns. The result indicates that Polyester Coolmax and Polyester gave the best results. Both of the fabrics gave the highest angle in both warp and weft. Polyester fiber is famously known for their wrinkle resistance property which is proven in the test.

Spun silk also gave a good result. However, the warp yarn prevents spun silk to be the best as it gives a low angle reading. Spun silk has less strength than filament silk which affected the tendency of the fabric to resist wrinkles.

Bamboo and Tencel yarns gave the worst result. This is because they are both cellulose base fiber and are known to have low wrinkle resistance properties. However, Bamboo yarn is better than Tencel yarn. This is because Bamboo yarn is much stronger and stiffer than Tencel yarn.

Drapability

Figure 7 shows the result of fabric drapability for each fabric with different type of weft yarn used. Good draping leads to the fitting of the fabric over a surface without undesirable wrinkling or tearing.

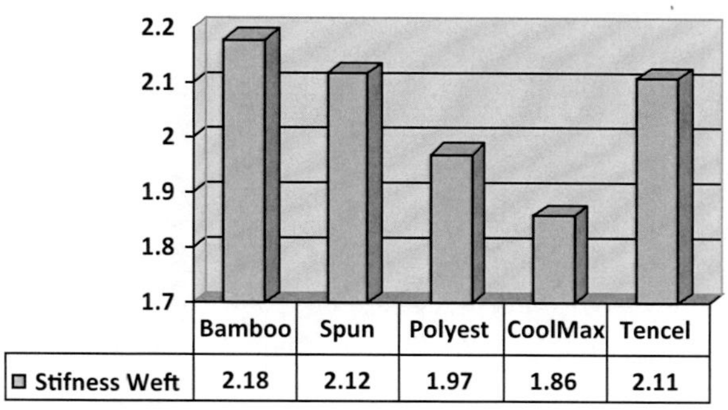

Fig. 8 Stiffness of the fabric

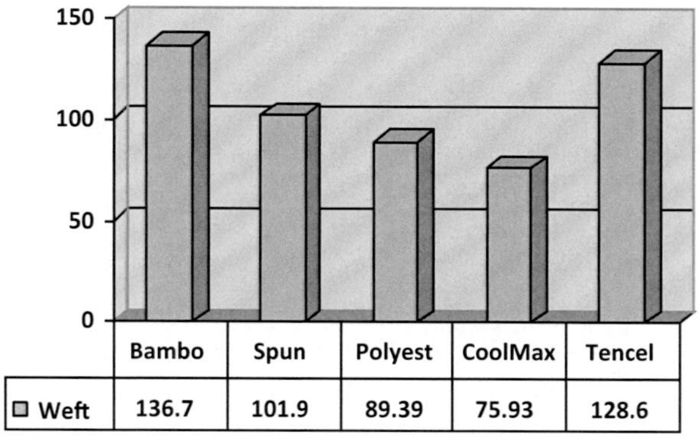

Fig. 9 Flexural rigidity

The results show that the Tencel fabric sample gave the best drapability compared to other fabric samples shown by the lowest drape coefficient, 47.58 %. As claimed by the fiber, manufacturer, Lenzing, the Tencel fibre was accepted to the textile industry because of its drapability.

On the other hand, the highest drape coefficient is the Bamboo fabric sample. Bamboo fabric sample was expected to have a poor drapability as the fabric is rather harsh and stiff.

Stiffness

Figure 8 shows the result of fabric stiffness for each fabric with different type of weft yarn used and Fig. 9 shows the flexural rigidity of the fabric. Fabrics that have a low bending rigidity were difficult to handle on an automated production line. The result shows that Bamboo has the highest flexural rigidity with 136.75.

The fabric needs more length to bend on its own weight. This is followed by Tencel yarns with 128.69. The fabric exceeds the expectation when the fabric was expected to be softer rather than stiff. This is followed by spun silk which is also famous for its softness. Polyester and Polyester Coolmax are also soft with flexural rigidity of 89.39 and 75.93, respectively.

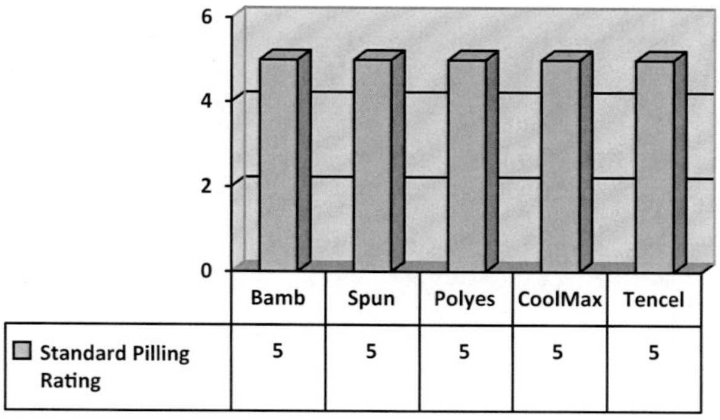

Fig. 10 Fabric pilling

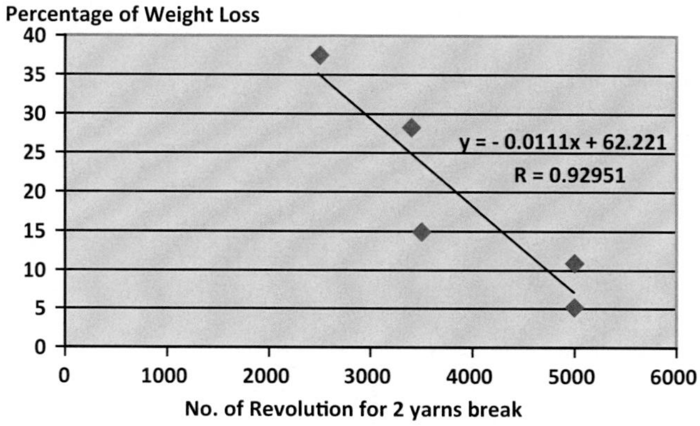

Fig. 11 The relationship between percentage of weight loss and number of cycles after 2 yarns break

Pilling

Figure 10 shows the average standard rating after 3600 revolutions of the pilling box. All of the samples show good result with rating of 5. Some of the fabric samples show a little pilling but can still be categorized as rating 5. This test is suitable if the fabric is not exposed to extreme rubbing.

Fabric Analysis

Relationship Between Weight Loss and the Number of Cycles

Figure 11 shows the relationship between the percentage of weight loss and the number of cycles after 2 yarns break. The results show a strong relationship between the percentage of weight loss and the number of cycles for 2 yarns break with a coefficient of R-value of 0.92.

The line shows that the number of revolution for 2 yarns break will affect the percentage of weight loss as the higher the number of revolution for 2 yarns break, the lower the

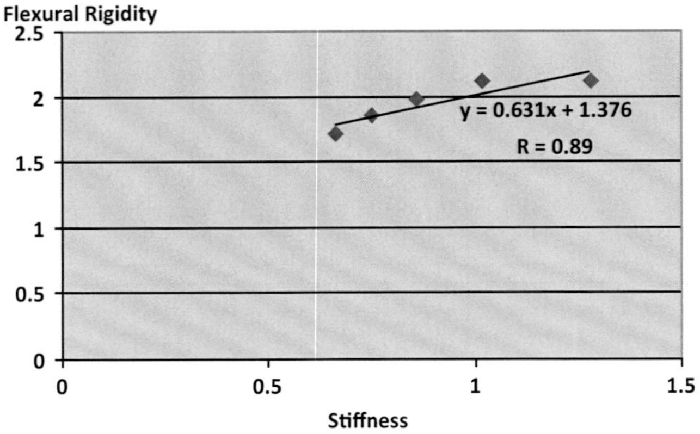

Fig. 12 The relationship between stiffness and flexural rigidity

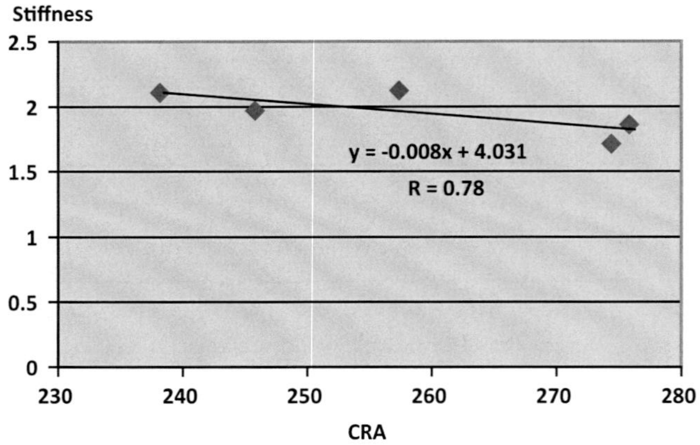

Fig. 13 The relationship between the crease angle recovery and the stiffness of the fabric

percentage of weight loss. Some fabrics need higher number of revolution for the yarns to break, thus the weight loss is less.

Relationship Between Stiffness and Flexural Rigidity

Figure 12 shows the relationship between the flexural rigidity and crease recovery angle. It can be seen that the relationship between the two factors is strong with R-value of 0.89. It can be said that the flexural rigidity is influenced by the crease recovery angle of the fabric.

The Relationship Between Crease Angle Recovery and Stiffness

Figure 13 shows the relationship between the crease recovery angle and stiffness indicating a strong R-value of 0.78. It can be said that the crease recovery angle is influenced by the stiffness of the fabric.

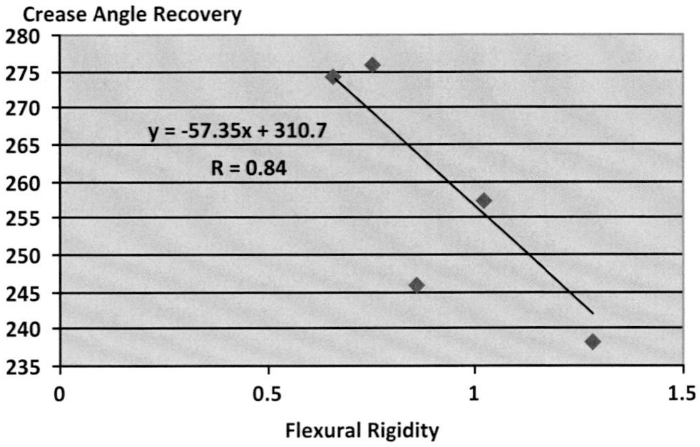

Fig. 14 The relationship between flexural rigidity and crease angle recovery

The Relationship Between Flexural Rigidity and Crease Angle Recovery

Figure 14 shows the relationship between the flexural rigidity and crease recovery angle. It can be seen that the relationship between the two factors is strong with R-value of 0.84. It suggests that the flexural rigidity is influenced by the crease recovery angle of the fabric.

Conclusion

From the results, it can be concluded that in terms of drapability, Tencel and Polyester Coolmax are comparable with spun silk, while in terms of abrasion, Polyester Coolmax shows the best result. In term of crease angle recovery, Polyester Coolmax and Bamboo show the best result. Overall, Polyester Coolmax, Tencel, and Bamboo have the potential to replace spun silk in Tenun Pahang.

Acknowledgment The authors would like to acknowledge the grant provided by the Ministry of Education through the Fundamental Research Grant Scheme (FRGS). The assistance from the Research Management Institutes (RMI) of Universiti Teknologi MARA is greatly appreciated.

References

1. H.F. Abdullah, S. Lecturer, Tenun Pahang Diraja **7**(1), 88–107 (2012)
2. B. Rafeah, *Kain Tenun Pahang* (Pahang, Lembaga Muzium Negeri Pahang, 2001)
3. Journey Malaysia. Kain Tenun (n.d). Retrieved 13 Aug 2013, from http://www.journeymalaysia.com/MCUL_kaintenun.htm
4. L.W. Kain Tenun (n.d). Retrieved 12 Aug 2013, from http://www.journeymalaysia.com/MCUL_kaintenun.htm
5. Perbadanan Kemajuan Negeri Pahang (n.d). Tenun Pahang. Retrieved 12 Aug 2013, from http://www.pknp.gov.my/web-v1/html/produk/usahawan/usahawan-07.php
6. Anonymous (n.d). Cited Oct 2012, from http://pkpahang.blogspot.com/p/institut-kemahiran-tenun-pahang-diraja.html

7. Arkib - Mstar (n.d). Cited Sept 2012, from http://mstar.com.my/cerita.asp?sec=mstar_rencana&file=/2012/5/24/mstar_rencana/20120524111753
8. D. Ma, Y. Yang, Application of New Environmentally Friendly Materials in Clothing. Design of Art, Art and Clothing College, Tianjin Polytechnic University, China, pp. 389–392 (n.d)
9. K. Dimitrovski, Z. Zupin, Mechanical properties of fabrics from cotton and biodegradable yarns bamboo, Spf, PLA in Weft. Faculty of Natural Sciences and Engineering, Department of Textiles, University of Ljubljana, Slovenia, pp. 25–45 (n.d)
10. K.S.N. Misfer, M.M.A. Haji, Y.M.E. Hassan, The Indian textile journal (2011). Retrieved 22 Nov 2013, from http://www.indiantextilejournal.com/articles
11. B. Wingate, *Textile Fabric and Their Selection* (Prentice-Hall, Englewood Cliffs, 1964)
12. M. Senthilkumar, B.B. Jambagi, Properties of spun silk knitted fabrics garments, fashion & retail features. Ind. Text. J. (2010). Department of Textile Technology. Retrieved 12 Aug 2013, from http://www.indiantextilejournal.com/articles/FAdetails.asp?id=3054
13. M.L. Cowan, *Introduction to Textile* (Appleton, New York, 1962)

Utilization of Eco-Colourant from Green Seaweed on Textile Dyeing

M.I. Ab Kadir, W.Y. Wan Ahmad, M.R. Ahmad, M.I. Misnon, W.S. Ruznan, H. Abdul Jabbar, K. Ngalib, and A. Ismail

Abstract

Synthetic dyes are toxic and harmful to the living things. Therefore, the demand for natural dyes is emerging globally due to the fact that they are safer and more environmental friendly, and thus, the application of natural dyes should be considered as a better alternative to synthetic dyes. Using natural dyes contributes to the acceptability of the product by the customers and also responses to the increasing demand of compatibility with the environment. Along the Malaysian coast, stretching along Peninsula, Sabah and Sarawak are the habitats for marine plants and algae. Previously, seaweeds are used for food, processed products and medicinal usage by local ethnics. However, this research focuses on the extraction of natural dye from green seaweed of *Caulerpa lentillifera* as textile colourant. *Caulerpa lentillifera* was extracted using boiling water and ammonia fermentation methods. The dyeing was then performed by exhaustion at 85 °C for 60 min. Three types of mordant were used by metachrome or simultaneous addition of mordant and dye in the dyebath. The dyed samples were then measured using spectrophotometer to analyse the shades obtained with regard to L*a*b* values and K/S values (colour strength). The dyed samples were also compared in terms of their ability to withstand washing, perspiration, rubbing/crocking and light. The results showed that the natural dye obtained from boiling water extraction method gave higher K/S values in comparison with the dye obtained from ammonia fermentation method. Fastness properties of the dyed samples were evaluated according to MS ISO standard and ranged from good to excellent rating except for lightfastness which is poor.

Keywords

Natural dyes • *Caulerpa lentillifera* • Extraction • Colour strength • Colourfastness

Introduction

Since prehistoric times until the mid- to late nineteenth century, natural dyes are the primary colour source for textiles [1]. However, the use of natural dyes declined rapidly after the discovery of synthetic dyes in 1856 by William Henry Perkin [2] because synthetic dyes are much cheaper, brighter and widely available [3].

Recently, the harmful effects of synthetic dyes to human being are becoming noticeable in which some of them are even carcinogenic. Natural dyes have many excellent properties such as less toxic, little side effect, biodegradable

Contract grant sponsor: Ministry of Education (MOE) under Exploratory Research Grant Scheme (ERGS)
Contract grant number: 600RxxMI/ERGS5/3/(11/2011)

M.I. Ab Kadir (✉) • W.Y. Wan Ahmad • M.R. Ahmad • M.I. Misnon • W.S. Ruznan • H. Abdul Jabbar
Faculty of Applied Sciences, Textile Research Centre, Universiti Teknologi MARA, 40450 Shah Alam, Selangor Darul Ehsan, Malaysia
e-mail: muhammad035@salam.uitm.edu.my

K. Ngalib • A. Ismail
Faculty of Applied Sciences, School of Biology, Universiti Teknologi MARA, 40450 Shah Alam, Selangor Darul Ehsan, Malaysia

M.R. Ahmad and M.F. Yahya (eds.), *Proceedings of the International Colloquium in Textile Engineering, Fashion, Apparel and Design 2014 (ICTEFAD 2014)*, DOI 10.1007/978-981-287-011-7_14,

Fig. 1 *Caulerpa lentillifera* seaweed

and eco-friendly as well as unique in shades [4]. Hence, the demand and interest in natural dyes have been increasing. Natural dyes are obtained from some plants (e.g. indigo and saffron), insects (e.g., cochineal and lac), animals (e.g. some species of molluscs or shellfish) and minerals (e.g. ferrous sulphate, ochre and clay) without any chemical treatment [5–7]. In addition, Balagurunathan et al. [8] claim that natural dye can also be extracted from microorganisms such as bacteria, fungi, algae and actinomycetes.

Seaweed refers to the large marine algae that grow almost exclusively in the shallow waters at the edge of the world's oceans. Seaweeds are plants because they use the sun's energy to produce carbohydrates from carbon dioxide and water (this is called photosynthesis). They are simpler than the land plants mainly because they absorb the nutrients that they require from the surrounding water and have no need for roots or complex conducting tissues [9]. There are three types of seaweeds that are classified by the pigments present, red (Rhodophyta), brown (Phaeophyta) and green (Chlorophyta) [9, 10]. Dennis [11] stated that the commercial exploitation of seaweeds is as food and the production of agar, alginate as well as carrageenan. Furthermore, seaweeds also exploited to be a source of bioethanol, dietary supplements, medicinal and pharmaceutical uses and cosmetics [9].

According to Phang [12], there are 375 taxa of seaweeds recorded in Malaysia. *Caulerpa lentillifera,* or also called as round sea grapes, as shown in Fig. 1 is a species of seaweed which is commonly known as green macro alga under the family of *Caulerpaceae*.

Materials and Methods

Materials

The *Caulerpa lentillifera* seaweeds were collected from Semporna, Sabah. 100 % plain weave silk fabric was used as the substrates. Two percent (2 %) of metallic salts of ferrous sulphate (iron) and potassium aluminium sulphate (alum) as well as acetic acid is used as mordants for each different dyebaths. Boiling water extraction and ammonia fermentation were performed to obtain colourants from *Caulerpa lentillifera.*

Experimental Methods

1. *Dye Extractions:* Two types of extraction methods were used. In the boiling water extraction method, a liquor ratio of 1:20 (weight of material in gram: amount of water in mL) was used to boil *Caulerpa lentillifera* in distilled water for 60 min. The extracted mixture was then cooled down and sieved before being used to dye the fabric. In the ammonia fermentation method, one part of ammonium hydroxide solution was diluted in 10 parts of distilled water. *Caulerpa lentillifera* was soaked in diluted ammonium hydroxide solution (same liquor ratio applied in boiling water extraction) at room temperature for 4 weeks [13]. The mixture was then sieved to obtain the extracts.
2. *Dyeing of Silk Fabric:* Silk was dyed with both extracted colourants using exhaustion dyeing technique. Two percent (2 %) colourant in the form of liquid, based on the weight of fabric, was used to dye the silk fabric using a liquor ratio of 1:20. About 2 % of each mordant was used to fix the colourant onto the fabric. Dyeing and mordanting were carried out simultaneously in one bath. The dyeing process was performed at 85 °C for 60 min. After the dyeing cycle was completed, the dyed fabrics were washed, rinsed with tap water and then left to dry.
3. *Colour Assessments:* The dyed samples were assessed in accordance to MS ISO standards. Colour fastness to washing, perspiration, rubbing/crocking and light was determined from standard test methods as listed in Table 1. The K/S value (colour strength) was measured using HunterLab LabScan XE (LSXE) spectrophotometer and analysed using HunterLab EasyMatchQC software.

Table 1 Standard methods used for colour fastness assessments

Colourfastness	Standard methods	Equipments
Washing	MS ISO 105-C01-1966	Auto-wash
	MS ISO 105-A05-2003	Change in colour
	MS ISO 105-A04-2003	Staining
Perspiration	MS ISO 105-E04-1996	Perspirometer
	MS ISO 105-A05-2003	Change in colour
	MS ISO 105-A04-2003	Staining
Rubbing/crocking	MS ISO 105-X12-2001	Crockmeter
	MS ISO 105-A04-2003	Staining
Light	MS ISO 105-B02-2001	Lightfastness tester

Results and Discussion

Dyed Samples

The swatches of dyed silk fabric with *Caulerpa lentillifera* are tabulated in Fig. 2. The shades obtained from both extraction methods are comparable.

L*a*b* Values

The L*a*b* values for silk fabric dyed with *Caulerpa lentillifera* extracts from boiling water extraction and ammonia fermentation methods are tabulated in Table 2. The L* values indicate perceived lightness or darkness. Value of 0 indicates black and 100 indicates white. The values of a* indicate red (+a) and green (−a), while b* indicates yellow (+b) and blue (−b). The darkest shades obtained from both extraction methods are from the fabric treated with iron as a mordant. However, the shades obtained from the boiling water extraction are darker than the shades obtained from ammonia fermentation. On the other hand, the lightest shades obtained from boiling water extraction are from the fabrics treated with alum as a mordant. However, for ammonia fermentation method, the treated fabric acetic acid and alum without mordant produced almost similar shades in terms of lightness. The coordinates for a*b* are plotted in 2D plots as shown in Figs. 3 and 4 accordingly.

Colour Strength

The K/S curves for all dyed samples are presented in Fig. 5. It is quite obvious that all the dyes show similar trends because they were from the same source. The colour strength (K/S) values were highest for silk dyed with dye from boiling water extraction and treated with iron as a mordant. In contrast, the value was lowest for silk dyed with dye from ammonia fermentation and treated with acetic acid.

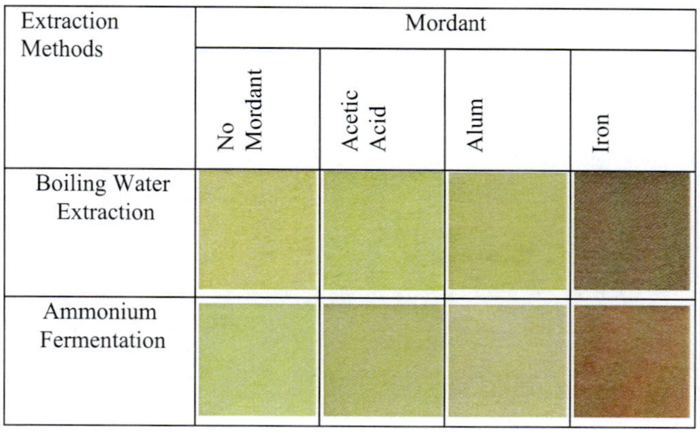

Fig. 2 Swatches of dyed silk

Table 2 L*a*b* values for dyed silk fabric with *Caulerpa lentillifera*

Extraction methods		Mordant	L*	a*	b*
Boiling water extraction	1	No mordant	74.49	−2.09	18.07
	2	Acetic acid	77.52	−1.75	18.67
	3	Alum	81.5	−1.19	18.12
	4	Iron	57.18	5.14	21.33
Ammonia fermentation	1	No mordant	85.64	0.49	15.04
	2	Acetic acid	85.35	0.37	13.27
	3	Alum	85.08	0.45	13.64
	4	Iron	66.92	13.26	29.23

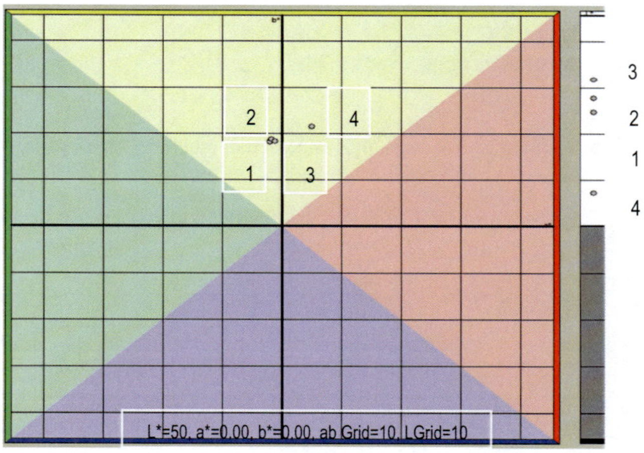

Fig. 3 2D plot represents the shades obtained from boiling water extraction

Generally, dyes extracted using boiling water gave higher K/S value after treated with mordant than dye extracted from ammonia fermentation.

Colourfastness Properties

Tables 3, 4 and 5 show the summary of fastness properties assessed from dyed silk fabric. Washing fastness for all dyed

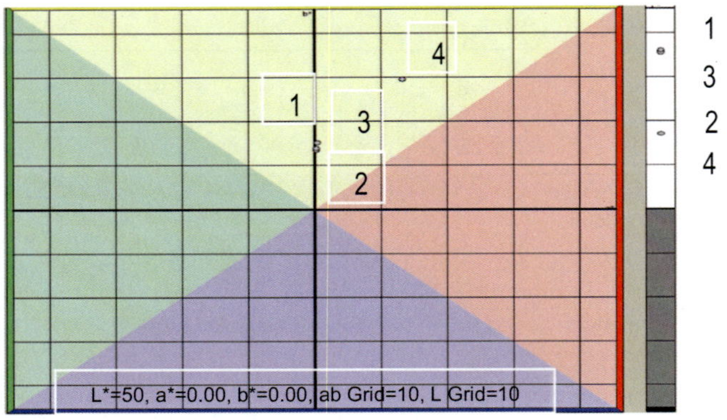

Fig. 4 2D plot represents the shades obtained from ammonia fermentation

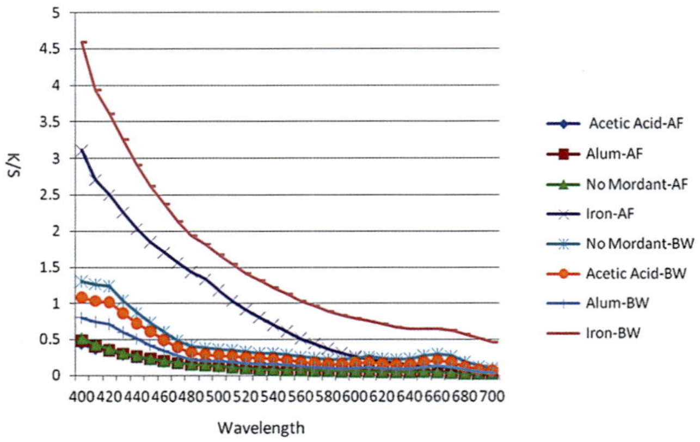

Fig. 5 K/S curves of the dyed samples

Table 3 Summary of washing fastness properties of the dyed silk fabric

		Washing		
			Staining	
Extraction methods	Mordants	Colour change	Cotton	Silk
Boiling water	No mordant	4	4/5	4/5
	Acetic acid	4	4/5	4/5
	Alum	4/5	4/5	4/5
	Iron	4	4/5	4/5
Ammonia fermentation	No mordant	4/5	5	5
	Acetic acid	4/5	4/5	4/5
	Alum	4/5	4/5	4/5
	Iron	4/5	4/5	4/5

samples were rated from 4 to 4/5 for change in colour and staining rated from 4/5 to 5. This rating is considered as good where 5 is the best. The result for fastness properties to perspiration also gave good rating of 4/5 for change in colour, and the rating for staining ranged from 4 to 4/5. Same goes to fastness properties result for rubbing/crocking which was from 3 to 5 for dry rub and ¾ to 4/5 for wet rub. Conversely, the fastness properties to light for all dyed samples were poor with rating of 3–4.

Table 4 Summary of fastness properties to perspiration of the dyed silk fabric

		Perspiration		
			Staining	
Extraction methods	Mordants	Colour change	Cotton	Silk
Boiling water	No mordant	4/5	4/5	4/5
	Acetic acid	4/5	4/5	4
	Alum	4/5	4/5	4
	Iron	4/5	4/5	4
Ammonia fermentation	No mordant	4/5	4/5	4/5
	Acetic acid	4/5	4/5	4/5
	Alum	4/5	4/5	4/5
	Iron	4/5	4/5	4/5

Table 5 Summary of fastness properties to rubbing/crocking and light of the dyed silk fabric

	Mordants	Rubbing/crocking		
Extraction methods		Dry	Wet	Light
Boiling water	No mordant	4/5	4	3
	Acetic acid	4/5	4	3
	Alum	4/5	4/5	3
	Iron	3/4	4/5	3
Ammonia fermentation	No mordant	4/5	4/5	4
	Acetic acid	5	4/5	4
	Alum	5	4/5	4
	Iron	3	3/4	4

Conclusion

From the study, it can be concluded that *Caulerpa lentillifera* seaweed can be exploited as a natural dye source which produces unique and interesting shades on textiles. The shades obtained vary from greenish yellow to reddish brown depending on the mordant used which gave acceptable fastness properties even without mordant.

Acknowledgement The authors acknowledge the financial support obtained from the Ministry of Education, Malaysia, (MOE) under the Exploratory Research Grant Scheme (ERGS). The assistance from the Research Management Institute (RMI) is also highly appreciated.

References

1. D. Cristea, I. Bareau, G. Vilarem, Identification and quantitative HPLC analysis of the main flavonoids present in weld (*Reseda Luteola* L.). Dyes Pigments **57**(3), 267–272 (2003)
2. D. Cristea, G. Vileram, Improving light fastness of natural dyes on cotton yarn. Dyes Pigments **70**, 238–245 (2006)
3. A.K. Samanta, P. Agarwal, Application of natural dyes on textiles. Ind. J. Fibre Text. Res. **34**, 384–399 (2009)
4. W.Y. Wan Ahmad, M.A. Mohd Nor, N. Saim, M.I. Ab Kadir, M.R. Ahmad, Nano natural dyes from *Melastoma Malabathricum* L. Adv. Mater. Res. **545**, 59–63 (2012)

5. T. Bechtold, A.M. Ali, R. Mussak, Natural dyes for textile dyeing: a comparison of methods to assess the quality of Canadian golden rod plant material. Dyes Pigments **75**, 287–293 (2006)
6. B. Chengaiah, R.K. Mallikarjuna, K.K. Mahesh, M. Alagusundaram, C.C. Madhusudhana, Medicinal importance of natural dyes: a review. Int. J. PharmTech Res. **2**(1), 144–154 (2010)
7. A. Manhita, T. Ferreira, A. Candeias, Extracting natural dyes from wool - an evaluation of extraction methods. Anal. Bioanal. Chem. **400**, 1501–1514 (2011)
8. R. Balagurunathan, T. Diraviyam, M. Radhakrishnan, Antioxidant activity of melanin pigment from *Streptomyces species* D5 isolated from Desert soil, Rajasthan, India. Drug Invent. Today **3**(3), 12–13 (2011)
9. K. Ajit, A.K. Meena, M.M. Rao, P. Panda, A.K. Mangal, G. Reddy, B. Ramesh, Marine algae: an introduction, food value and medicinal uses. J. Pharm. Res. **4**(1), 219–221 (2011)
10. C.S. Lobban, M.J. Wynne, The biology of seaweeds. Bot. Monogr. **17**, 786 (1981)
11. J.M. Dennis, Worldwide distribution of commercial resources of seaweeds including *Gelidium*. Hydrobiologia **221**, 19–29 (1991)
12. S.M. Phang, Seaweed resources in Malaysia: current status and future prospects. Aquat. Ecosyst. Health Manag. **9**(2), 185–202 (2006)
13. S.G. Brough, Navajo lichen dyes. Lichenologist **20**(3), 279–290 (1988)

Dyeing of Polyester Using Natural Colorant from *Melastoma malabathricum* L.

Wan Yunus Wan Ahmad, Tengku Muna Shaheera Tuan Zainal Abidin, Mohd Rozi Ahmad, Muhammad Ismail Ab Kadir, and Nor Juliana Mohd Yusof

Abstract

Nowadays, interests on natural colorant have increased as the public are more aware and concern to environmental-related issues. In the continuing replacement of synthetic colorant, applications of natural colorant have extended to synthetic fibres. In this study, *Melastoma malabathricum* L. was extracted through boiling and ultrasound-assisted extraction (UAE). The colorant from UAE was dried in an oven before being converted into powder form using ball mill grinder. The colorant produced was dyed on polyester fabric at 85 °C. The polyester fabric was treated with sodium hydroxide as an alkaline treatment and chitosan prior to dyeing in an attempt to increase the dye uptake. The evaluation of the dyed fabrics was measured using colour spectrophotometer and tested for washfastness. Samples dyed using the UAE method gave lower value of L* which indicate higher dye uptake and darker colour. The fabric which was treated with sodium hydroxide gave deeper shade in comparison with the fabric treated with chitosan. The ratings for colourfastness to washing were between 3 and 5 which show moderate to good range.

Keywords

Natural colorant • Polyester • Ultrasound-assisted extraction

Introduction

Colour is one of characteristic of textiles for eye-appeal and even for aesthetic purposes. Synthetic fibres are synonym with the application of synthetic colorant. However, in recent years, a considerable amount of attention has been generated for the use of natural colorant on dyeing synthetic fibres [1]. Natural colorant has been normally known for their use in colouring natural protein fibres like wool, silk and cellulose fibres like cotton as major areas of application since pre-historic times [2].

Before the advent of the synthetic colorant in second half of nineteenth century, all colorants were derived directly from vegetables or animal sources [3]. There are plenty of plants that can be the source for natural colorant such as pomegranate [4], henna leaves [5] and marigold flower [6]. The *Melastoma malabathricum* L. plant, locally known as 'senduduk', is one of the potential natural colorants. This plant is a common weed that grows abundantly in waste grounds and open fields in Malaysia. Fruits from this plant have deep purplish colour that can be a source of natural colorant. Different parts of the plants have been utilized for the treatment of several diseases in folk medicine [7].

The conventional processes of natural dyes are by mechanical ways like grinding, crushing and steeping in water to extract the coloured liquid. However, researchers are trying to develop more appropriate techniques to produce more effective and efficient extraction of available matters from the plant materials. Ultrasound is one of the techniques

W.Y. Wan Ahmad (✉) • T.M.S. Tuan Zainal Abidin •
M.R. Ahmad • M.I. Ab Kadir • N.J. Mohd Yusof
Faculty of Applied Sciences, Textile Research Center, Universiti Teknologi MARA, 40450 Shah Alam, Selangor, Malaysia
e-mail: wanyunus@salam.uitm.edu.my

M.R. Ahmad and M.F. Yahya (eds.), *Proceedings of the International Colloquium in Textile Engineering, Fashion, Apparel and Design 2014 (ICTEFAD 2014)*, DOI 10.1007/978-981-287-011-7_15,

that is found to be more effective in saving energy and also use of moderate temperatures [8]. In view of the fact that colouring matter is a solid–liquid leaching process which can involve mass transfer problem, extraction could be more improved by ultrasound method. This novel technique can improve the main mechanism extraction of natural dyes such as rupture of the cell wall and release of natural dyes as well as advanced the transport of dye into the external medium. Venkatasubramanian also claimed that the used of ultrasound is found to have significant improvement in the extraction efficiency of colorant obtained from beetroot and is beneficial in natural dyeing of leather with improved rate of exhaustion [9].

The study conducted here was to apply natural colorant from *Melastoma malabathricum* L. on polyester fabric at dyeing temperature lower than 100 °C. Polyester is one of the synthetic fibres that posses high durability and resistant to wrinkle. However, the fibre is hydrophobic and non-absorbent without chemical modification. Due to its hydrophobicity and high crystallinity, polyester is difficult to dye. Since polyester have no active dye-site, it is impossible to be dyed with majority of dyestuff [10]. Normally, polyester is dyed using disperse dyes at temperature of 130 °C. However in order to overcome the deficiency in dyeing of polyester, natural colorant from *Melastoma malabathricum* L. was extracted using ultrasound and converted into powder in small sizes to ensure that the colorant was able to penetrate into polyester polymer system. In addition, to facilitate the dyeing process, the polyester fabric was treated with sodium hydroxide (alkaline treatment) and chitosan to improve the affinity between natural dye and polyester fabric.

Materials and Methods

Materials

This research used plain woven polyester fabrics with an areal density of 95 g/m^2. Sodium hydroxide was supplied by Merck, Germany. Acetic acid and Chitosan purchased from R&M Marketing, United Kingdom and *Melastoma malabathricum* L. was collected at Bukit Bandaraya, Shah Alam Selangor. All chemicals were used without further purification.

Methods

Fabric Treatment

The polyester fabric was treated with alkaline solution and chitosan treatment to increase the dye uptake during the dyeing process. In the alkaline treatment, the fabric was treated with 10 % sodium hydroxide for 2 h at 60 °C. As for the chitosan treatment, 3 g of chitosan was dispersed in 1 L aqueous acetic acid for every 5 g fabric [11]. The concentration for acetic acid was 1 % and the samples were agitated for 2 h at 60 °C.

Extraction and Dyeing

Extraction of natural dye from *Melastoma malabathricum* L. was carried out through boiling and ultrasound-assisted extraction (UAE). *Melastoma malabathricum* L. fruits were soaked into distilled water and heated for an hour at 100 °C. The dye solution was then filtered and dyed on polyester fabric. On the other hand, UAE was carried out at lower temperature of 50 °C. The ratio of fruits and distilled water in ultrasound-assisted extraction was 1:5. The ultrasound probe was dipped into the samples with frequencies of 20 kHz for 3 h. The extracted natural dye was dried in an oven and grinded using ball mill to produce powdered natural dye. The natural dye from *Melastoma malabathricum* L. was applied to polyester fabric at 85 °C for an hour.

Colour Measurements and Colourfastness Evaluation

The CIELab value and 2D Plot were measured using ColourQuest XE spectrophotometer under illuminant D_{65} and 10^0 standard observer. The L*, a* and b* values represent lightness (L*), redness (+a*), greenness (−a*), yellowness (+b*) and blueness (−b*).

The dyed fabric was also analysed according to colourfastness to washing using combination of MS ISO standard methods of MS ISO 105-C01-1996, MS ISO 105-A05-2003 and MS ISO 105-A04-2003.

Result and Discussion

Cie L* a* b* Value and 2D Plot

Cie L* a* b* value was measured on dyed polyester fabrics using colour spectrophotometer.

Table 1 and Fig. 1 show Cie L* a* b* value and 2D Plot of dyed polyester using *Melastoma malabatricum* L. extracted through boiling and ultrasound-assisted extraction.

Table 1 Cie L* a* b* value of *Melastoma malabathricum* L. on dyed fabrics

Sample	L*	a*	b*
BE untreated	74.68	4.41	−10.74
BE alkaline	73.27	4.85	−10.35
BE chitosan	75.75	2.81	−2.32
US untreated	72.05	3.82	−11.69
US alkaline	69.16	3.73	−11.28
US chitosan	76.04	2.23	−0.55

BE Boiling extraction, *US* Ultrasound-assisted extraction

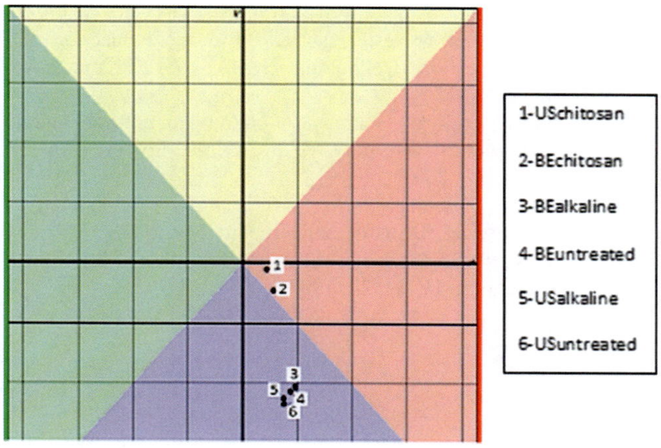

Fig. 1 2D plot represents the shades on dyed polyester

Ultrasound improved extraction process in comparison to boiling method. Samples treated with sodium hydroxide gave lower L* compared to untreated samples, even though the samples fall under the same shade area as shown in Fig. 1 of 2D plot. Lower L* means darker colour shade which indicates higher dye uptake achieved due to the possible increase in the number of hydrophilic group on the fibre surface caused by chain scission after the fabric was treated with alkaline [12]. Samples treated with chitosan produced light colour when dyed with *Melastoma malabatricum* L. A related study conducted by Saxena et al. [13] on dye uptake of lac dye on chitosan-treated cotton showed that chitosan enhanced dye uptake and at the same time increased dye sorption. However in this case where polyester was treated with chitosan, the dyeing effect produced was very minimal. Figure 1 showed that polyester treated with Chitosan gave bluish red colour.

Colourfastness Testing

Colourfastness to washing rate is from 1 to 5, where 5 indicates the best result and 1 is the worst. This rating showed the ability of the dye to retain their colour on the fabric after washing process. Table 2 shows the rating of colour change and staining for dyed polyester fabrics using *Melastoma malabatricum* L. The rating for colour change was between 3 and 4 which means moderate changes occurred after washing process. The staining on polyester and cotton were 4–5, which is good. Alkaline treatment samples gave moderate colour change compared to untreated and chitosan-treated samples in ultrasound-assisted extraction.

Table 2 Colourfastness to washing test on dyed polyester fabric

		Washing		
			Staining	
Sample	Extraction	Colour change	Cotton	Polyester
a	Boiling	3/4	5	5
	Ultrasound	3	4/5	5
b	Boiling	4	5	5
	Ultrasound	3	4/5	5
c	Boiling	4	5	5
	Ultrasound	4	4/5	5

(a) Untreated sample
(b) Alkaline treatment sample
(c) Chitosan treatment sample

Conclusion

The research of *Melastoma malabatricum* L. natural dye on polyester showed that polyester fabric can be dyed at low temperature of 85 °C. This can definitely save energy. At the same time alkaline treatment on polyester fabric is able to improve the absorption of dye thus increasing the dye uptake which resulted in darker shades. However treatment of polyester with chitosan did not improve the dye ability of polyester using *Melastoma malabatricum* L., but the colourfastness rating to washing was good.

Acknowledgement The authors acknowledge the financial support obtained from Ministry of Higher Education (MOHE) under the Research Grant Acculturation Scheme (RAGS). The assistance from the Research Management Institute (RMI) and Faculty of Applied Sciences are also highly appreciated.

References

1. H. Lokhande, V.A. Dorugade, Dyeing nylon with natural dyes.pdf. Am. Dyestuff Rep., 29–34 (1999)
2. A. Kumar, P. Agarwal, Application of natural dyes on textiles. 34 (December), 384–399 (2009)
3. C.H. Giles, *A Laboratory Course in Dyeing*, 2nd edn. (Chorley & Pickersgill, Leeds, 1971)
4. S.S. Kulkarni, A.V. Gokhale, U.M. Bodake, G.R. Pathade, Cotton dyeing with natural dye extracted from pomegranate (*Punica granatum*) peel. Univ. J. Environ. Res. Technol. **1**(2), 135–139 (2011)
5. M.M. Alam, M.L. Rahman, M.Z. Haque, Extraction of henna leaf dye and its dyeing effects on textile fibre. Bangladesh J. Sci. Ind. Res. **42**(2), 217–222 (2007)
6. D. Jothi, Extraction of natural dyes from african marigold flower (*Tagetes ereecta L*) for textile coloration. AUTEX Res. J. **8**(June), 49–53 (2008)

7. M.R. Sulaiman, M.N. Somchit, D.A. Israf, Antinociceptive effect of *Melastoma malabathricum* ethanolic extract in mice. Fitoterapia **75**, 667–672 (2004)
8. M.D. Esclapez, A. Mulet, J.A. Ca, Ultrasound-assisted extraction of natural products. Food Eng. Rev. **3**, 108–120 (2011)
9. G.S. Venkatasubramanian Sivakumar, J. Lakshmi Anna, J. Vijayeeswarri, Ultrasound assisted enhancement in natural dye extraction from beetroot for industrial applications and natural dyeing of leather. Ultrason. Sonochem. **16**, 782–789 (2009)
10. M. Ismail, W. Yunus, W. Ahmad, M.R. Ahmad, M.I. Misnon, H.A. Jabbar, Fastness properties and colorimetric characteristics of low temperature dyeing of natural dyes from the barks of *Ixonanthes icosandra* jack on polyester fabric. Business Engineering and Industrial Applications Colloquium (BEIAC) (IEEE, Langkawi, 2013), pp. 448–452
11. S.H.A. Shadi Houshyar, Treatment of cotton with chitosan and its effect on dyeability with reactive dyes. Iran. Polym. J. **11**(5), 295–301 (2012)
12. M.J.C.S. Haig Zeronian, *Surface Modification of Polyester by Alkaline Treatments* (British Textile Technology, 1989)
13. S. Saxena, V. Iyer, A.I. Shaikh, V.A. Shenal, Dyeing of cotton with lac dye. Colourage **44**(11), 23 (1997)

Microwave-Enzyme-Assisted Extraction and Dyeing of Lichen Species: *Parmotrema praesorediosum*

N.A. Mohamed, W.Y. Wan Ahmad, K. Ngalib, M.R. Ahmad, M.I. Ab Kadir, and A. Ismail

Abstract

This research work involves the dyeing of silk fabrics with lichen *Parmotrema praesorediosum*. The dyes were extracted using two techniques involving microwave-assisted extraction (MAE) and microwave-enzyme-assisted extraction (MEAE). This was found to be comparable with conventional boiling technique. The CIE L*a*b* and K/S values of the dyed samples were measured. Colourfastness properties including washing, rubbing and light of each sample was measured. The research results indicated that modification of MEAE led to higher dye absorption shade compared to MAE and boiling method. The optimal microwave power is 700 W. The colourfastness testing shows that all samples produced good colourfastness for natural dye.

Keywords

Natural dye • Lichen • Microwave assisted • Enzyme • Colourfastness

Introduction

Natural dyes are obtained from natural sources, most commonly from vegetables, plants, trees and a few from insects. The shades produced by natural dyes are usually soft, lustrous and soothing to the human eye. In fact, natural dyes are believed to be safe due to its non-toxic, non-carcinogenic and biodegradable nature [1].

Lichens are very common organisms which can survive in a wide variety of environments, including on rocks, trees and other suitable substrates. Lichen belongs to the kingdom Protista and Fungi as it is a symbiosis made up either of a fungal partner or an algal partner or both [2]. More than 20,000 species of lichens have been identified [3]. Lichen has the greatest capabilities of adapting themselves to the most conditions of climate, altitude, moisture, heat and cold. The wide diversity of lichen flora in Malaysia has been published by Galloway et al. [4] to provide general perspective on the study of lichen in Malaysia. There are reports on the studies of Malaysian lichen [5, 6]; to date, only limited report is found on the natural colourant of Malaysian lichens [7].

Lichen dyes are known as substantive dyes which mean that it does not require a mordant to help them fix dyes to the fibres. Lichen contains parietin which is the predominant cortical pigment of lichens and responsible for brightly colour appearance of lichen which can be extracted for dyestuff [8]. Extraction of lichen dyes was demonstrated in a study conducted by Casselman [9]. The most common colour yields from lichen are red, brown, yellow and blue pigments depending on the different species of lichen, and these pigments have been used as textile dyes since the ancient time. Native Americans used boiled lichen extracts to dye cloth and baskets. Even the famous Harris Tweeds, woollen textiles from the Scottish islands, are still dyed with Scottish lichens.

There is growing demand for developing suitable extraction techniques for more effective and efficient

N.A. Mohamed • W.Y. Wan Ahmad (✉) • M.R. Ahmad • M.I. Ab Kadir
Faculty of Applied Sciences, Textile Research Centre, Universiti Teknologi MARA, 40450 Shah Alam, Selangor Darul Ehsan, Malaysia
e-mail: wanyunus@salam.uitm.edu.my

K. Ngalib • A. Ismail
Faculty of Applied Sciences, School of Biology, Universiti Teknologi MARA, 40450 Shah Alam, Selangor Darul Ehsan, Malaysia

M.R. Ahmad and M.F. Yahya (eds.), *Proceedings of the International Colloquium in Textile Engineering, Fashion, Apparel and Design 2014 (ICTEFAD 2014)*, DOI 10.1007/978-981-287-011-7_16,

extraction of plant materials. Started in the late 1970s, the use of microwave energy as a heating source in analytical laboratories was applied [10]. Indeed, the novel process such as the microwave-assisted extraction (MAE) has been studied over conventional extraction methods. MAE possesses the advantages of considerable reduction in time. There were reports on the preparation of natural dye from cochineal by MAE [11]. A study on MAE of natural red pigment from *Pitaya* [12] showed that MAE method is recommendable to the application for extraction of natural colourants.

In line with MAE, enzymes have been considerable use for many years in textile industry [13, 14]. Roughly, amylases are used in desizing, cellulases in denim finishing and bio-polishing of cellulosic fibres and proteases in leather, silk and wool processing. Enzymes operate in mild conditions and are biodegradable. Various enzyme combinations are used to loosen the structural integrity of botanical material, thereby enhancing the extraction of desired flavour and colour components [15]. Previous study by Padma et al. [16] had used enzymes for dyeing cotton with natural dye. They found that enzymatic treatment gave cotton and silk fabric rapid dye absorption and offer an alternative to the metal-mordanted natural dyeing.

Therefore, this study identifies the potential use of lichen dyes, and comparison was made between dyeing with lichen extracts from microwave-assisted extraction (MAE) and microwave-enzyme-assisted extraction (MEAE). The colour coordinates and the depth of shades of all samples were evaluated in terms of CIELAB and K/S values of the dyed substrate.

Materials and Methods

Materials

Satin weave silk fabric (71 g/m^2) was used as the substrate. Lichen species of *Parmotrema praesorediosum* was collected from Cameron Highland, Pahang, Malaysia. The lichen was thoroughly washed, dried and shredded into small pieces before the extraction process.

Extraction Methods

Boiling Extraction

The shredded lichen was dissolved in distilled water with a liquor ratio of 1:40 and allowed to soak for 60 min. The mixture was boiled for another 60 min. The content was cooled off and filtered before being used as dyeing solution.

Microwave-Assisted Extraction (MAE)

MAE was performed with a domestic microwave oven (maximal power: 700 W; Samsung, China). The mixture was placed in a screw-cap bottle and allowed to soak for 60 min. The bottle containing the sample was then placed into the microwave oven, and the extraction was carried out for 5 min (power settings at 180, 450 and 700 W). After irradiation, the extraction solutions were cooled and filtered before being used as dyeing solution.

Microwave-Enzyme-Assisted Extraction (MEAE)

A 2 % solution of pectinase to cellulose (2:1) was sprayed on lichen and then left overnight. After that, the enzyme-treated lichen extraction was carried out similar with the MAE as mentioned in section 'Microwave-Assisted Extraction (MAE)' [17].

Dyeing Procedure

Silk samples were dyed using the lichen-extracted dye solution at 50 % of weight fabric (owf) concentration with a liquor ratio of 1:40. Dyeing was done in a laboratory exhaustion dyeing machine, Labtec, at 85 °C for 60 min. After dyeing, the dyed fabrics were rinsed and air-dried.

Colour Measurements

The CIELAB values and colour strength (K/S) values of the dyed samples were measured using ColorQuest XE spectrophotometer under illuminant D_{65}, 10^0 standard observer. The CIE L*a*b* colour system utilizes 3 coordinates to locate a colour in colour space. L* indicates lightness and can have values between 0 and100, $a^* > 0$ redness, $a^* < 0$ greenness, $b^* > 0$ yellowness and $b^* < 0$ blueness.

Through the L* values, colour can be classified as light or dark. The higher the number of L*, the lighter the colour, and the lower the number, the more dull the colour, whereas a* and b* values for the dyed samples that intersect at the colour space will identify, respectively, as points in the 2D plot. These points specify the hue (colour) and chroma (vividness/dullness). K/S is the absorption coefficient (K) to scattering coefficient (S) giving reflectance values at wavelengths between 400 and 700 nm.

Colourfastness Evaluation

The dyed samples were tested for their colourfastness to washing according to MS ISO 105-C01-1966, MS ISO 105-A05-2003 and MS ISO 105-A04-2003. Colourfastness to rubbing was done according to MS ISO 105-X12-2001 and MS ISO 105-A04-2003. Wash, dry and wet rubbing fastness change in the shade was evaluated using the AATCC Gray Scale for Evaluating Change in Colour, and colour transfer was evaluated using the AATCC Gray Scale for Evaluating Staining. Colourfastness to light testing was carried out following the MS ISO 105-B02-2001 and was assessed using the blue wool scale.

Results and Discussion

Colour Analysis

The colour samples of silk dyed fabrics using lichen *P. Praesorediosum* are shown in Fig. 1. The dyed samples exhibited different shades of beige, brown and brownish yellow. Therefore, this dyestuff achieved to give a uniform dyeing results and capable to be used for dyeing silk fabric.

The CIE L*a*b* values for silk dyed with lichen extracts from boiling, MAE and MEAE methods are tabulated in Table 1. The lowest L* value achieved was 76.18, observed in sample dyed with lichen extracts from MEAE-700W. The other sample dyed with lichen extracts from MAE-700W exhibited L* value of 77.36. This is the second lowest value found. Boiling extraction exhibited a fairly lower L* value of 77.51. Therefore, the sample dyed with lichen extracts from MEAE-700W had the deepest colour appearance, whereas the sample dyed with lichen extracts from MAE-700W and boiling had the lightest colour, based on L* values. It was found that the enzyme assisted the MAE method to improve the extraction of these extracted dyes. Enzymes have the ability to degrade or disrupt cell walls and membranes, thus enabling better release and more efficient extraction of bioactive [18].

Similar trends can be observed by other samples dyed from MEAE-450W which exhibited L* value lower than sample dyed from MAE-450W. The same thing can be said for the L* value of MEAE-180W which was lower than MAE-180W. The results also suggested that the extraction efficiency improved by raising the microwave power from 180 to 700 W. It was found that during the experiment, the increase in the microwave power increases the dye yield [19]. MAE heats the extract quickly and accelerates the extraction process of targeted compounds from the matrix [20]. MAE can save a lot of time compared to boiling method and gave higher dye absorption.

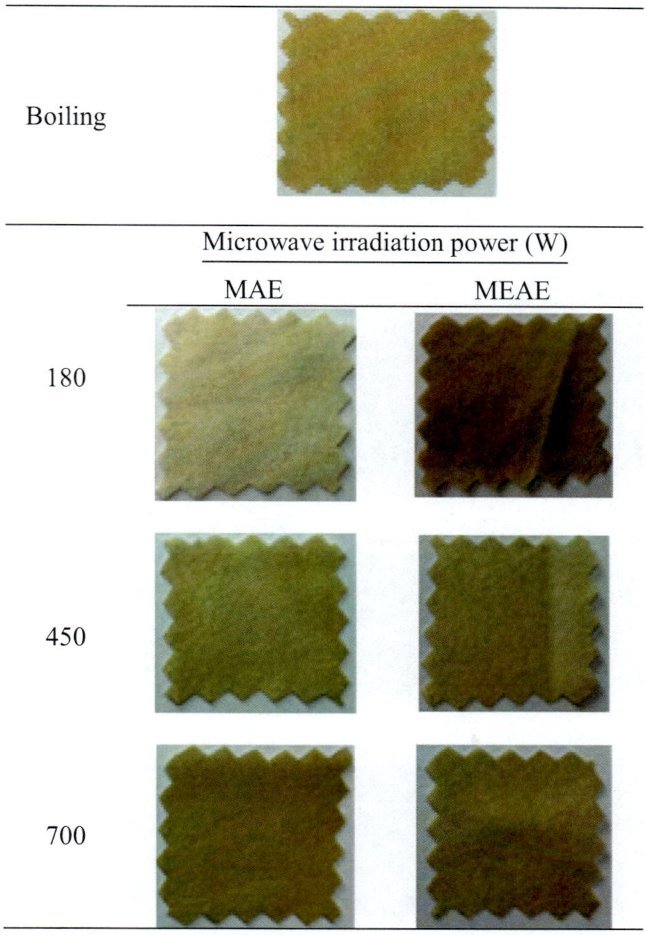

Fig. 1 Silk dyed samples

Table 1 CIELAB values for silk dyed with P *praesorediosum*

Methods	L*	a*	b*
Boiling	77.51	4.47	18.48
MAE-180W	81.82	1.81	11.16
MAE-450W	77.73	3.1	15.58
MAE-700W	77.36	2.81	15.07
MEAE-180W	77.93	2.54	12.07
MEAE-450W	77.72	2.69	11.78
MEAE-700W	76.18	2.69	11.98

The coordinates for L*a*b* are also plotted in 2D plots as shown in Fig. 2. The variables of a* values of samples dyed with all extraction methods are positive, indicating colour on the red side along the red-green axis, whereas the change in the b* values is positive showing yellow colour along the yellow-blue axis.

The K/S curves of all dyed samples are presented in Fig. 3. It is shown that all dyed samples exhibited a similar trend due to its origin from the same source. In general, from the plots of curves, it shows that the extracted dyes with

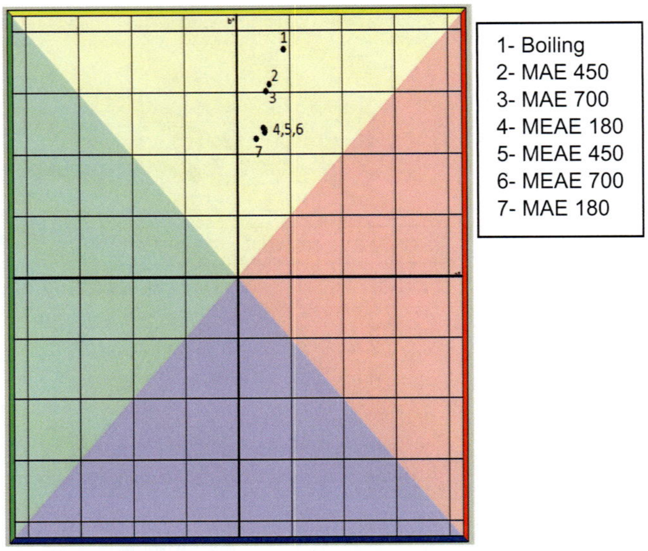

Fig. 2 2D plot represents the shades obtained from dyed silk fabrics

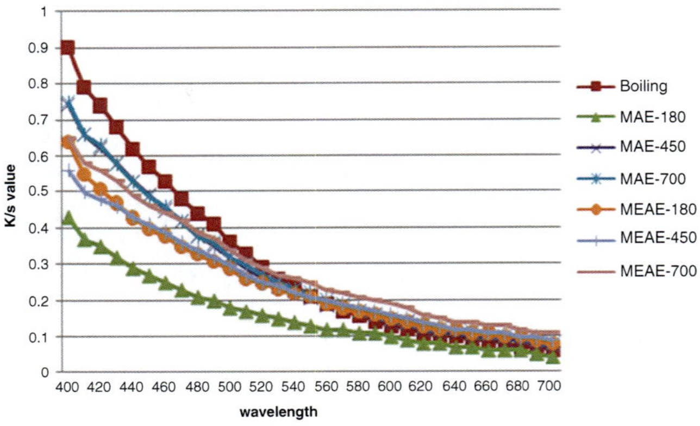

Fig. 3 Plots of absorption curves measured from silk dyed fabrics

MEAE gave higher dye absorption, increased in the order of MEAE-700W > MAE-700W > boiling > MEAE-450W > MAE-450W > MEAE-180W > MAE-180.

Colourfastness Properties

Table 2 lists the colourfastness properties of the dyed silk fabric. Visually for wash fastness and rubbing testing, there was no significant fading of the colour, and so the greyscale rating from 4 to 5 was assessed as good to excellent.

As for lightfastness results, the fabric was assessed by comparing with blue wool standards. The rating ranges from 1 to 8, where 1 is the lowest and 8 is the best. All dyed samples were rated as 4. The results indicated reasonable lightfastness for a natural dye. Liu et al. suggested that to improve lightfastness, the dyeing process can be conducted by adding mordant [21].

Table 2 Fastness properties of dyed silk fabric

	Washing			Rubbing/crocking		
		Staining				
Extraction	Colour change	Silk	Cotton	Dry	Wet	Light
Boiling	4	4	4	5	5	4
MAE-180	4	4	4	5	5	4
MAE-450	4	4	4/5	5	5	4
MAE-700	4	4	4	5	5	4
MEAE-180	4	4	4	5	5	4
MEAE-450	4	4	4/5	5	5	4
MEAE-700	4	4	4	5	5	4

Conclusions

Dye extraction from lichen *P. Praesorediosum* was successfully carried out by all three methods. Generally, silk dyed with lichen extracts from MEAE gave darkest appearance according to L* values and exhibited higher dye absorption. The second higher dye absorption is obtained by MAE. The optimal microwave irradiation power was 700 W. Colourfastness properties of all dyed samples obtained good to excellent with a greyscale rating from 4 to 5. It can be concluded that MAE has high potential as an alternative to conventional extraction together with the use of enzyme-assisted extraction (MEAE). In this case, the use of MAE is preferable on the basis of fast heating, efficient and energy saving. This study also shows that enzyme is the ideal catalyst to assist in the extraction. Enzyme helps in reducing solvent usage and is biodegradable and inexpensive. All these factors make both MAEA and MAE extremely interesting from the point of view of an industry that strives to reduce their environmental problem.

Acknowledgement The authors would like to acknowledge the Ministry of Education, Malaysia (MOE), for the Fundamental Research Grant (FRGS) and the assistance from the Research Management Institute (RMI), Universiti Teknologi MARA, Malaysia. The authors are also thankful to Prof. Dato' Dr. Mohd Wahid Samsudin and Prof. Dato' Dr. Laily Bin Din from the Faculty of Science and Technology, Universiti Kebangsaan, Malaysia, for their assistance with lichen identification.

References

1. S.J. Kadolph, K.D. Casselman, In the bag: contact natural dyes. Cloth. Text. Res. J. **22**(1–2), 15–21 (2004)
2. H.N. Thomas, *Lichen Biology*, 2nd edn. (Cambridge University Press, Cambridge, 2008)
3. D.J. Galloway, M.W. Samsudin, A. Latiff, A bibliography of Malaysian lichenology. Malaysian Appl. Biol. **22**, 215–221 (1994)
4. D.J. Galloway, L.B. Din, A. Latiff, An additional bibliography of Malaysian lichenology. Malaysian Appl. Biol. **28**, 93–99 (1997)

5. L.B. Din, Z. Zakaria, M.W. Samsudin, J.A. Elix, Chemical profile of compounds from lichens of Bukit Larut, Peninsular Malaysia. Sains Malaysiana **39**(6), 901–908 (2010)
6. E.L. Din, G. Ismail, J.A. Elix, The lichens in Bario Highlands: their natural occurrence and secondary metabolites. Biodivers. Environ. Conserv. **1**(6), 1 (1999)
7. N.A. Mohamed, W.Y.W. Ahmad, K. Ngalib, M.R. Ahmad, Dyeing of silk fabric with natural dyes from lichen using ultrasound assisted extraction, in *IEEE Symposium on Humanities, Science and Engineering Research*, 2013, pp. 4–7
8. J.G. Romagni, F.E. Dayan, Structural diversity of lichen metabolites and their potential use. in *Advances in Microbial Toxin Research and Its Biotechnological Exploitation* (Springer, New York, 2002), pp. 151–158
9. K.D. Casselman, *Lichen Dyes: The New Source Book* (Dover, Mineola, NY, 2001)
10. B. Kaufmann, P. Christen, Recent extraction techniques for natural products: microwave-assisted extraction and pressurised solvent extraction. Phytochem. Anal. **13**(2), 105–13 (2002)
11. Y. Guo, H. Zheng, H. Zhang, L. Ma, J. Han, K. Li, Optimization of combined microwave-ultrasonic wave extraction of cochineal dye by response surface methodology. Appl. Mech. Mater. **161**, 82–87 (2012)
12. L. Changhui, Y. Xiandong, W. Xiaorong, H. Shenggen, H. Pengcheng, Z. Weifeng, Study on microwave-assisted extraction of natural red pigment from pericarp of pitaya. J. Hunan Agric. Univ. Nat. Sci. **6** (2009)
13. E. Tsatsaroni, M. Liakopoulou-Kyriakides, Effects of enzymatic treatment on the dyeing of cotton and wool fibres with natural dyes. Dyes Pigments **29**(3), 203–209 (1995)
14. M. Liakopoulou-Kyriakides, E. Tsatsaroni, P. Laderos, K. Georgiadou, Dyeing of cotton and wool fibres with pigments from *Crocus sativus*-effect of enzymatic treatment. Dyes Pigments **36**(3), 215–221 (1998)
15. H.B. Sowbhagya, V.N. Chitra, Enzyme-assisted extraction of flavorings and colorants from plant materials. Food Sci. Nutr. **50**(2), 146–161 (2010)
16. P.S. Vankar, R. Shanker, A. Verma, Enzymatic natural dyeing of cotton and silk fabrics without metal mordants. J. Clean. Prod. **15**(15), 1441–1450 (2007)
17. H.C. Tiwari, P. Singh, P.K. Mishra, P. Srivastava, Evaluation of various techniques for extraction of natural colorants from pomegranate rind-ultrasound and enzyme assisted extraction. Fibers Text. Res. **35**, 272–276 (2009)
18. M. Puri, D. Sharma, J.B. Colin, Enzyme-assisted extraction of bioactives from plants. Trends Biotechnol. 8 (2011)
19. I.T. Karabegovi, S.S. Stoji, D.T. Veli, N.Č. Nikoli, M.L. Lazi, A. Materials, Optimization of microwave-assisted extraction of cherry laurel (*Prunus laurocerasus L.*) fruit using response surface methodology, 351–356 (2012)
20. Z. Lianfu, L. Zelong, Optimization and comparison of ultrasound/microwave assisted extraction (UMAE) and ultrasonic assisted extraction (UAE) of lycopene from tomatoes. Ultrason. Sonochem. **15**, 731–737 (2008)
21. Y. Liu, G. Bai, Bioecological mordant dyeing of silk with monascorubrin. Text. Res. J. **82**(2), 203–207 (2011)

Microwave-Assisted Extraction as a Rapid Extraction to Produce Natural Dyes from *Pycnoporus sanguineus* Mushroom

W.Y. Wan Ahmad, N. Md Noor, M.R. Ahmad, and M.I. Ab Kadir

Abstract

The demand for natural dyes has prompted researchers to look for modern techniques to extract the natural dyes from plant and fungi. In this study, domestic microwave oven was used as a rapid extraction method to extract dyes from mushroom from the *Pycnoporus sanguineus* species. The power of the microwave oven was set at 300, 450 and 600 W. The time of exposure with the microwave radiation was 5 min. For comparison, extraction was also conducted using conventional extraction method. Dyeing was then performed by exhaustion method at 80 °C dyeing temperature for 60 min. The dyed samples were measured using spectrophotometer to analyse the shades obtained with regards to their L*a*b* values. The dyed samples were also compared in terms of their colourfastness performance to withstand washing, perspiration, rubbing/crocking and light.

Keywords

Pycnoporus sanguineus • Extraction • Boiling • Microwave assisted • Colourfastness

Introduction

Mushroom is an organism from the kingdom of Fungi that is known for their nutritional value and the diversity of their bioactive component [1]. These kinds of fungi can be used in various applications such as in the field of food technology, medical and agriculture. However, not all of the species can be consumed by human, not only because of the poisonous elements contained in the mushroom but also because of their appearance, taste and texture [2]. Based on the fact that some of the inedible mushrooms are not harmful to human, they can be explored as a potential source of natural dyes. The first mushroom dyes was discovered by Miriam C. Rice from California [3] when she began experimenting natural dyes to make her own ink for her block printing.

Natural resources can be considered as very important sources of natural dyes. They have become a part of the history and identity of different cultures [4]. Man, with creativity in mind, had utilised materials such as mineral, plants and animal origin and convert them into dyes [5]. In general, natural dyes are extracted using boiling water or by fermentation methods. However, these conventional methods present several drawbacks such as long extraction time, potential loss of volatile constituents and requirement of high energy use [6]. Thus, developing an alternative extraction technique that is rapid, sensitive, safe and energy efficient is highly desirable. New extraction techniques have been applied such as microwave-assisted extraction (MAE), supercritical fluid extraction (SFE) and ultrasonic-assisted extraction (UAE) to shorten time, reduce organic solvent consumption, improve extraction yield, enhance extract quality, prevent pollution and reduce sample preparation cost [7–9].

Microwave (MW) is one of the components of electromagnetic energy in the range of frequency between 300 MHz and 300 GHz or between wavelengths of 1 cm and 1 m [10]. Microwaves are used as an information carrier and to absorb a part of the electromagnetic energy from materials and transform to heat. The use of microwave

W.Y. Wan Ahmad • N. Md Noor (✉) • M.R. Ahmad • M.I. Ab Kadir
Faculty of Applied Sciences, Textile Research Center, Universiti Teknologi MARA, 40450 Shah Alam, Selangor Darul Ehsan, Malaysia
e-mail: norasidahmdnoor@gmail.com

M.R. Ahmad and M.F. Yahya (eds.), *Proceedings of the International Colloquium in Textile Engineering, Fashion, Apparel and Design 2014 (ICTEFAD 2014)*, DOI 10.1007/978-981-287-011-7_17,

energy as a heating source in analytical laboratories started in the late 1970s, and the first microwave-assisted extraction was reported by Ganzler et al. in 1986 [7]. The normal frequency for commercial application in domestic oven is 2,450 MHz [11].

Materials and Methods

Materials

Mushrooms from the species of *Pycnoporus sanguineus* as shown in Fig. 1 were collected from Tasik Kenyir, Kuala Terengganu, Malaysia. The mushrooms were then washed and oven dried at 40 °C ± 2 °C for 36 h. The dried samples were ground and kept in a dry place. A 100 % satin weave silk with a weight of 71 g/m^2 was used as the substrate.

Conventional Boiling Water Extraction (CBWE)

In this method, 25 g of dried mushroom was weighed in a round bottom flask and distilled water was added using liquor ratio of 1:20 (weight of material in gram: amount of water in mL) [12]. The mixtures were left for 45 min before the extraction process. Extraction was carried out at 100 °C for 60 min on the hot plate. The mouth of the flask was covered using aluminium foil to reduce the evaporation process.

Microwave-Assisted Extraction (MAE)

Approximately 25 g of dried mushroom was weighed in a round bottom flask and distilled water was added using liquor ratio of 1:20. The mixtures were left for 45 min before the extraction process. The flask was exposed for 1 min in the microwave oven and the mixtures were cooled down in the oven for another 1 min. The duration of exposure and duration of cooling were done alternately, and the steps were repeated until the total duration of exposure to microwave radiation was added up to 5 min [13]. The exposure process was repeated using the power of 300, 450 and 600 W respectively.

Dyeing of Fabric

Dyeing was carried out using the exhaustion method onto silk fabric using liquor ratio of 1:20 at 80 °C dyeing temperature for 60 min. After thorough rinsing and drying, the dyed samples were measured to establish colour coordinates using Hunter Lab Scan XE spectrophotometer and computer. The fastness properties of the dyed fabrics were also conducted.

Fig. 1 *Pycnoporus sanguineus*

Table 1 Standard methods used for colourfastness assessments

Colourfastness	Standard method	Equipment
Washing	MS ISO 105-C01-1966	Auto-wash
	MS ISO 105-A05-2003	Change in colour
	MS ISO 105-A04-2003	Staining
Perspiration	MS ISO 105-E04-1996	Perspirometer
	MS ISO 105-A05-2003	Change in colour
	MS ISO 105-A04-2003	Staining
Rubbing/crocking	MS ISO 105-X12-2001	Crockmeter
	MS ISO 105-A04-2003	Staining
Light	MS ISO 105-B02-2001	Lightfastness tester

Fastness Testing

The dyed silk fabric was tested according to the standard methods as shown in Table 1.

Results and Discussion

L*a*b* Values

The dyed samples are shown in Fig. 2, while the L*a*b* values for silk fabric dyed with the *Pycnoporus sanguineus* mushroom are given in Table 2. The L* value indicates perceived lightness or darkness where value of 0 indicates black and 100 indicates white. The value of a* indicates red (+a) and green (−a), while b* indicates yellow (+b) and blue (−b).

Based on Table 2, fabrics dyed using the MAE method gave darker shades in comparison to fabrics dyed using the boiling method. It can be said that the fabric absorbed more dyes with the MAE method. The MAE method can be used as alternative to extract mushroom dyes because the method can extract more dye particles in a short time (10 min) as compared to BWE (60 min). However, the amount of dyes

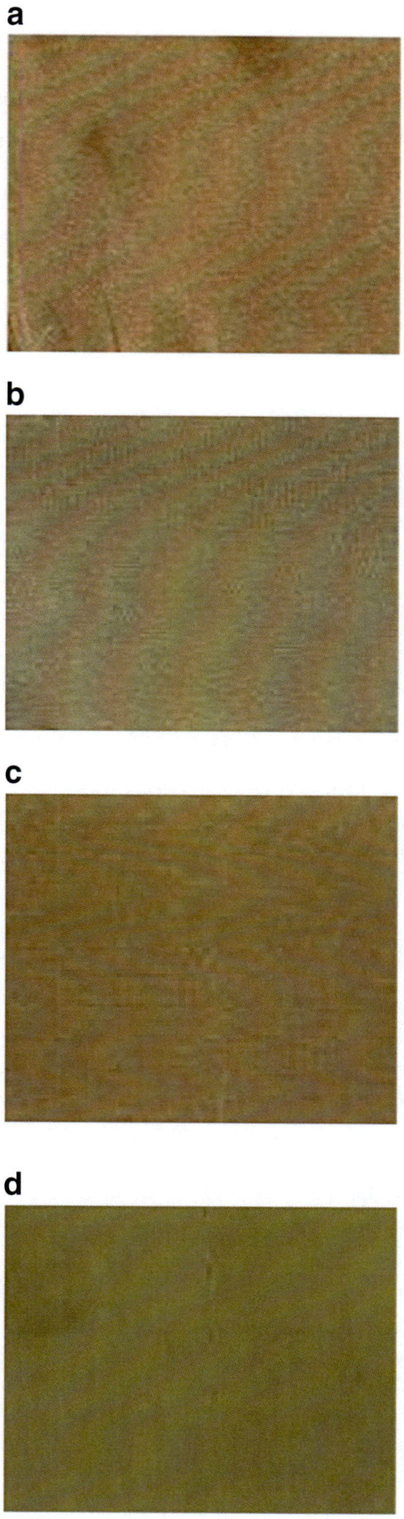

Fig. 2 BWE mixtures and MAE mixtures on silk: (**a**) BWE, (**b**) MAE-300W, (**c**) MAE-450W, (**d**) MAE-600W

particle extracted was less. This may be because the CBWE depends on conduction–convection phenomenon where most of the heat energy are lost to the environment, whereas in the MAE, heating occurs in a confined and closed area with practically no heat loss to the environment [10]. Therefore, the microwaves cause the glandular walls of the material to quickly rupture, resulting in high extraction efficiency at a shorter time. Besides that, the microwave power also played an important role in the extraction of the *Pycnoporus sanguineous* dyes. Based on Table 2, the power of 600 W indicates higher absorption of dyes compared to 300 W and 450 W. This is due to the drastically increment of the desorption of compound from the active sites in the matrix [6].

Table 2 L*a*b* values for dyed silk fabric with the mushrooms (*Pycnoporus sanguineus)* dye

Extraction methods	Power (Watts)	L*	a*	b*
Boiling water	–	76.09	6.81	31.25
Microwave assisted	300	78.17	4.91	28.70
	450	77.61	4.48	36.19
	600	74.96	7.51	35.28

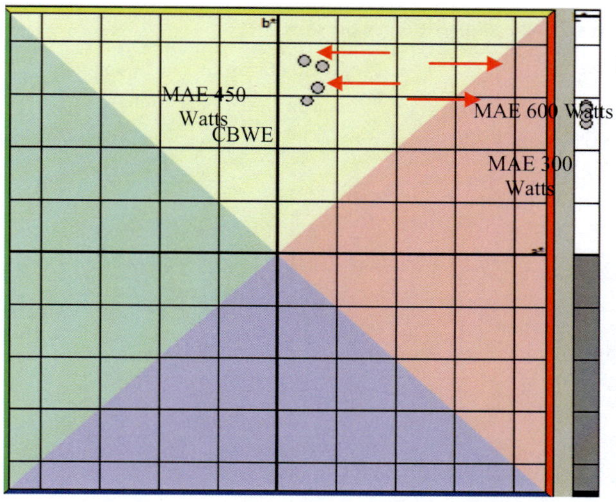

Fig. 3 2D plot represents the shades obtained from MAE mixture and BWE mixture

Figure 3 shows the 2D plot the shade of the dyed sample. All the samples are in the yellow region of the dye spectrum with influence of the reddish shade.

Colourfastness Properties

Wash fastness results for dyed silk as shown in Table 3 indicate that the change in colour are moderate to good as the rating is from 3 to 4. The same results were obtained from colourfastness to perspiration (Table 4) which gave rating from 3 to 3/4. The light fastness properties as shown in Table 5 indicate that these dyes have moderate light fastness with a rating of 3. This is due to photo oxidation

Table 3 Colourfastness to washing

		Colourfastness to washing		
		Change in	Staining on	
Extracted dyes		colour	Silk	Cotton
Boiling water		3	3/4	4
Microwave-assisted extraction (power–watts)	300	3	3	3/4
	450	3/4	3/4	3/4
	600	3/4	3	3/4

Table 4 Colourfastness to perspiration

		Colourfastness to perspiration		
		Change in	Staining on	
Extracted dyes		colour	Silk	Cotton
Boiling water		3	2/3	2/3
Microwave-assisted extraction (power–watts)	300	3	3	3/4
	450	3/4	3/4	3/4
	600	3/4	3	3/4

Table 5 Colourfastness to rubbing and light

		Colourfastness to rubbing		Colour fastness to light
Extracted dyes		Dry	Wet	
Boiling water		4	4	3
Microwave-assisted extraction (power–watts)	300	4	4	3
	450	4	3/4	3
	600	3/4	3/4	3

of the chromophore and poor bonding of the dyes molecules in the fibres as they are exposed to the light. In addition, colourfastness to rubbing gave good result with rating 3/4 to 4. All of the above rating is in agreement with natural dyes extracted from plant material [14].

Conclusion

The experiment using MAE indicated that it is an attractive alternative to conventional extraction such as CBWE and fermentation. The main advantage of MAE is in the performance of the heating source. The high temperature attained by microwave heating dramatically reduces the extraction time. Furthermore, the shades of the dyes can be changed if we manipulate the time of extraction.

MAE has a high potential to be used as a rapid extraction to extract natural dyes from mushroom such as the species of *Pycnoporus sanguineus* which depends on the power of radiation and time of exposure. In addition, the MAE method only takes 5 min or less to get the desired shades compared to the boiling extraction method. In this sense, the shades can be dictated depending on the time of exposure and power. This could be the next target of research.

References

1. J. Erjavec, J. Kos, M. Ravnikar, T. Dreo, J. Sabotic, Protein of higher fungi-from forest to application. Trends Biotechnol. Rev. **30** (5), 259 (2012)
2. J.M. Polese, *The Pocket Guide to Mushrooms* (Konemann Verlagsgesellschaft mbH, Cologne, 2000)
3. M.B. Dorothy, A brief history of the art of mushrooms dyeing. http://www.sonic.net/~dbeebee/web_mush-history.htm. Retrieved 19 Apr 2012
4. P.S. Vankar, Chemistry of natural dyes. Resonance, 73–80 (2002)
5. M. Cedano, L. Villasenor, L. Guzman-Davalos, Some aphyllophorales tested for organic dyes. Mycologist **15**(2), 81–85 (2002)
6. W.W. Hong, Q.L. Yan, L.W. Shou, J.Y. Zi, L. Kuan, Comparison of microwave-assisted and conventional in the extraction of essential oils from mango (*Mangifera indica* L.) flowers. Molecules **1**(5), 7715–7723 (2010)
7. B. Kaufmann, P. Christen, Recent extraction techniques for natural product: microwave-assisted extraction and pressurised solvent extraction. Phytochem. Anal. **13**, 105–113 (2002)
8. S. Hemwimol, P. Pavasant, A. Shotipruk, Ultrasound-assisted extraction of anthraquinones from roots of *Morinda citrifolia*. Ultrason. Sonochem. **13**, 243–548 (2006)
9. M.E. Borges, R.L. Tejera, L. Diaz, P. Esparza, E. Ibanes, Natural dyes extraction from cochineal (*Dactylopius coccus*). New extraction method. Food Chem. **132**, 1855–1860 (2012)
10. H. Ramaswamy, J. Tang, Microwave and radio frequency heating. Food Sci. Technol. Int. **14**(5), 423–427 (2009)
11. T. Jain, V. Jain, R. Pandey, A. Vyas, S.S. Shukla, Microwave assisted extraction for phytoconstituents: an review. Asian J. Res. Chem. **2**(1), 19–25 (2009)
12. W.Y. Wan Ahmad, M.A. Mohd Nor, N. Saim, M.I. Ab Kadir, M.R. Ahmad, Nano natural dyes from *Melastoma malabathrium* L. Adv. Mater. Res. **545**, 59–63 (2012)
13. V. Mandal, Y. Mohan, S. Hemalatha, Microwave assisted extraction-an innovative and promising extraction tool for medical plant research. Pharmacogn. Rev. **1**(1), 7–18 (2007)
14. R. Rahim, W.Y. Wan Ahmad, M.R. Ahmad, M.I. Ab Kadir, M.I. Misnon, The application of Gluta aptera wood (Rengas) as natural dye on silk and cotton fabrics. Univ. J. Environ. Res. Technol. **1**(4), 545–551 (2011)

Dyeing Properties and Absorption Study of Natural Dyes from Seaweeds, *Kappaphycus alvarezii*

M.I. Ab Kadir, W.Y. Wan Ahmad, M.R. Ahmad, H. Abdul Jabbar, K. Ngalib, and A. Ismail

Abstract

This research focuses on the extraction of natural dyes from red, green and brown strains of *Kappaphycus alvarezii* for textile colourant. The green, brown and red strains of *Kappaphycus alvarezii* seaweeds were extracted using boiling water and microwave-assisted extraction methods. The extracted colourant was measured with UV–Vis spectrophotometer to analyse the absorbance value at the maximum wavelength or peak absorbance (λ_{max}) as well as to determine the compounds present which contribute to the shades obtained. The dyeing was then performed by exhaustion at 85 °C for 60 min. Alum and iron were used as mordant and applied simultaneously in the dyebath. The dyed samples were then measured using spectrophotometer to analyse the shades obtained with regard to L*a*b* values and reflectance curves. The dyed samples were also compared in terms of their ability to withstand washing, perspiration, rubbing/crocking and light. The results showed that natural dye obtained from microwave-assisted extraction method gave darker shades in comparison with the dye obtained from boiling water extraction. Fastness properties of the dyed samples were evaluated according to MS ISO standard. The results ranged from good to excellent rating except for lightfastness which gave poor rating.

Keywords

Natural dyes • *Kappaphycus alvarezii* • Extraction • Maximum absorption peaks • Colourfastness

Introduction

The global textile industries used synthetic dyes extensively due to their availability, reproducibility, full shades of gamut, intensity in colours, being easy to use, better fastness properties and being less expensive. However, due to the toxic nature and adverse effects of synthetic dyes on environment as well as living things, more researches have been conducted substantially to explore the hidden components found in natural sources which could be exploited as an alternative to synthetic dyes. Recently, the subject of natural dyes has become the matter of topical interest and back in demand because natural dyes produce very unique, multihues, uncommon, soothing and soft shades as well as produce eco-performing textiles in comparison with synthetic dyes [1–3]. Thus, extensive researches are conducted from time to time to improve colour yield, explore new potential and sustainable sources, look for simpler extraction procedures as well as improve dyeability of the fibres. In fact, Shishoo [4] stressed that in order to remain competitive in the near future, the textile processing industries must practise

Contract grant sponsor: Ministry of Education (MOE) under Exploratory Research Grant Scheme (ERGS); contract grant number: 600RMI/ERGS5/3/(11/2011)

M.I. Ab Kadir (✉) • W.Y. Wan Ahmad • M.R. Ahmad • H. Abdul Jabbar
Faculty of Applied Sciences, Textile Research Center, Universiti Teknologi MARA, 40450 Shah Alam, Selangor Darul Ehsan, Malaysia
e-mail: muhammad035@salam.uitm.edu.my

K. Ngalib • A. Ismail
Faculty of Applied Sciences, School of Biology, Universiti Teknologi MARA, 40450 Shah Alam, Selangor Darul Ehsan, Malaysia

M.R. Ahmad and M.F. Yahya (eds.), *Proceedings of the International Colloquium in Textile Engineering, Fashion, Apparel and Design 2014 (ICTEFAD 2014)*, DOI 10.1007/978-981-287-011-7_18,

sustainable technologies and develop environmentally safer processing methods.

Traditionally, natural dyes are obtained through tedious extraction procedures from natural sources such as plants (e.g. indigo and woad), insects and invertebrates (e.g. cochineal, kermes and some species of molluscs) as well as minerals (e.g. ferrous sulphate, ochre and clay) [5]. These conventional extraction procedures, i.e. boiling extraction, would produce very low colouring component, e.g. about 1,200 molluscs are needed to produce 14 g of dye [6, 7]. Lately, extensive researches have successfully worked on extracting natural colourants from microorganisms such as bacteria, fungi, algae and actinomycetes [8–10].

Seaweeds are macroalgae under the kingdom Protista. These primitive photosynthetic multicellular eukaryotic marine algae are classified into three broad groups based on their pigment constituents [11]. Red algae (Rhodophyta) consists of 6,500 species having pigments identified as r-phycoerythrin and c-phycocyanin pigments, 3,000 species of brown algae (Phaeophyta) with pigments known as xanthophylls and fucoxanthins and some 7,000 species of green algae (Chlorophyta) with chlorophylls a and b, carotenes and various xanthophylls [12]. Red and brown algae are almost exclusively marine, but only 900 species out of 7,000 species of green algae are marine [12, 13].

Kappaphycus are red seaweeds (Rhodophyta). There are currently seven species of *Kappaphycus* worldwide [14]. They can be found in East Africa, Indonesia, Malaysia, Micronesia and the Philippines. One of the species, known as *K. alverezii*, produces different strains due to the different pigments in their thallus [15].

Seaweeds are the essential part of billions of industries in the world with their variety of usage in food, pharmaceuticals, phycocolloids, agriculture as well as a feedstock for biofuel. Singh and Gu [16], Hughes et al. [17], Arad and Yaron [18] and Prasanna et al. [19] stated that algae have a wide variety of natural pigments like chlorophyll, carotenoids and phycobiliproteins, which exhibit colours ranging from green, yellow, brown and red. Algae pigments have great commercial value as natural colourants in nutraceutical, cosmetics and pharmaceutical industry as well as health benefits [20, 21].

Materials and Methods

Materials

Red seaweeds (Rhodophyta) of *K. alvarezii* species formally known as *Eucheuma cottonii* (red, green and brown strains) as shown in Figs. 1, 2 and 3 and labelled as S1, S2 and S3 accordingly were collected from Bum Bum Island, Semporna, Sabah (4°27′27″N 118°41′23″E), Malaysia.

Fig. 1 Red strain of *Kappaphycus alvarezii* (S1)

Fig. 2 Green strain of *Kappaphycus alvarezii* (S2)

Fig. 3 Brown strain of *Kappaphycus alvarezii* (S3)

Hundred percent of satin weave silk and plain weave bamboo fabrics were used as substrates. Two percent (2 %) of metallic salts of ferrous sulphate (iron) and potassium aluminium sulphate (alum) were used as mordants for each different dyebaths. Boiling water extraction and microwave-assisted extraction were performed to obtain colourants from the *K. alvarezii* species.

Experimental Methods

1. *Dye Extractions:* Two types of extraction methods were used, i.e. boiling water extraction (BWE) and microwave-assisted extraction (MAE). In the boiling water extraction method, a liquor ratio of 1:20 (weight of material in gram/amount of water in mL) was used to boil *K. alvarezii* species in distilled water for 60 min. In the microwave-assisted extraction, the same liquor ratio was used, but the heating time used was 5 min at two different heating powers which were 450 and 700 W. The microwave oven used was Samsung MW71B and the frequency was set at 2,450 MHz. After the extraction is completed, the mixture in the beaker is then allowed to cool down to room temperature followed by filtration process sieved before being used to dye the fabrics.
2. *Dyeing of Fabrics:* Silk and bamboo fabrics were dyed with both extracted colourants using exhaustion dyeing technique. Four percent (4 %) of colourants in the form of liquid, based on weight of fabric, were used to dye silk fabric using a liquor ratio of 1:20. About 2 % of each mordant was used to fix the colourant onto the fabric. Dyeing and mordanting were carried out simultaneously in one bath. The dyeing process was performed at 85 °C for 60 min. After the dyeing cycle was completed, the dyed fabrics were washed, rinsed with tap water and then left to dry.
3. *Colour Assessments:* The dyed fabrics were assessed in accordance to MS ISO standards as tabulated in Table 1. Colour fastness to washing, perspiration, rubbing/crocking and light were determined from standard test methods as listed in Table 1. The K/S value (colour strength) was measured using HunterLab LabScan XE (LSXE) spectrophotometer and analysed using HunterLab EasyMatchQC software.
4. *Spectrophotometric Analysis:* The extracted colourants from red, green and brown strains of *K. alvarezii* were analysed using a Shimadzu UV–Vis spectrophotometer UV-2101. The UV–Vis spectrums of the colouring matter from each strain were obtained in the visible range of 350–800 nm. Then the pigment compounds' presence in the extracted solution was determined by measuring the absorbance value at the maximum wavelength or peak absorbance (λ_{max}).

Table 1 Standard methods used for colourfastness assessments

Colour fastness	Standard methods	Equipments
Washing	MS ISO 105-C01-1966	Auto-wash
	MS ISO 105-A05-2003	Change in colour
	MS ISO 105-A04-2003	Staining
Perspiration	MS ISO 105-E04-1996	Perspirometer
	MS ISO 105-A05-2003	Change in colour
	MS ISO 105-A04-2003	Staining
Rubbing/crocking	MS ISO 105-X12-2001	Crockmeter
	MS ISO 105-A04-2003	Staining
Light	MS ISO 105-B02-2001	Light fastness tester

Table 2 L* a* b* values for S1

Dyed samples	L*	a*	b*
S1BBUM	90.4	−3.23	18.36
S1BBA	92.66	−3.95	20.54
S1BBI	85.16	1.76	20.14
S1BSUM	90.77	−3.83	18.44
S1BSA	90.72	−0.21	10.97
S1BSI	78.61	3.34	17.51
S1MLBUM	91.61	−0.16	5.29
S1MLBA	92.1	−0.2	5.8
S1MLBI	85.87	2.92	18.04
S1MLSUM	91.48	−0.31	7.45
S1MLSA	91.91	0.27	7.88
S1MLSI	78.94	3.84	17.64
S1MHBUM	91.19	−0.38	6.66
S1MHBA	91.62	−0.4	5.8
S1MHBI	84.91	1.9	16.32
S1MHSUM	90.54	0.08	8.66
S1MHSA	90.45	0.44	8.94
S1MHSI	77.8	3.13	17.1

Results and Discussion

Colourimetric Coordinates of the Dyed Fabrics

The L*a*b* values for bamboo and silk fabrics dyed with red, green and brown strains of *K. alvarezii* species extract from boiling water extraction and microwave-assisted extraction methods are tabulated in Tables 2, 3 and 4, respectively. The L* values indicate perceived lightness or darkness. Value of 0 indicates black and 100 indicates white. The values of a* indicate red (+a) and green (−a), while b* indicate yellow (+b) and blue (−b).

Generally, the L* values for all dyed fabrics were reduced (which indicates darker shades) when treated with iron as mordant as shown in Figs. 4, 5 and 6. The darkest shades obtained from S1 were achieved from high power (700 W) MAE which L* values coordinated at 84.91 and 77.8 for both bamboo and silk fabrics accordingly. For S2, the

Table 3 L* a* b* values for S2

Dyed samples	L*	a*	b*
S2BBUM	93.48	1.22	−3.02
S2BBA	91.98	−0.51	7.07
S2BBI	85.42	2.13	17.01
S2BSUM	91.32	1.32	3.17
S2BSA	91.27	0.39	9.77
S2BSI	79.47	2.82	14.86
S2MLBUM	91.42	−0.42	5.35
S2MLBA	92.03	−0.56	6.09
S2MLBI	85.34	2.17	16.99
S2MLSUM	90.6	0.03	6.61
S2MLSA	88.1	0.31	7.3
S2MLSI	78.44	3.17	18.48
S2MHBUM	90.81	−0.57	5.5
S2MHBA	91.64	−0.56	5.24
S2MHBI	90.2	−0.37	5.49
S2MHSUM	90.17	−0.06	7.68
S2MHSA	89.54	0.65	10.39
S2MHSI	89.13	0.66	9.15

Table 4 L* a* b* values for S3

Dyed samples	L*	a*	b*
S3BBUM	90.89	−1.5	11.16
S3BBA	92.02	−1.45	10.08
S3BBI	85.81	2.07	18.14
S3BSUM	90.58	−1.56	12.1
S3BSA	91.63	−0.16	9.5
S3BSI	78.13	3.16	15.55
S3MLBUM	91.6	−0.15	4.94
S3MLBA	92.18	−0.4	5.87
S3MLBI	85.71	2.1	16.7
S3MLSUM	91.31	0.15	7.96
S3MLSA	89.64	0.92	10.21
S3MLSI	77.9	3.8	17.28
S3MHBUM	91.73	−0.52	5.37
S3MHBA	91.3	−0.46	6
S3MHBI	86.19	1.92	16.83
S3MHSUM	89.85	0.04	9.38
S3MHSA	89.88	0.72	9.25
S3MHSI	78.23	3.34	15.85

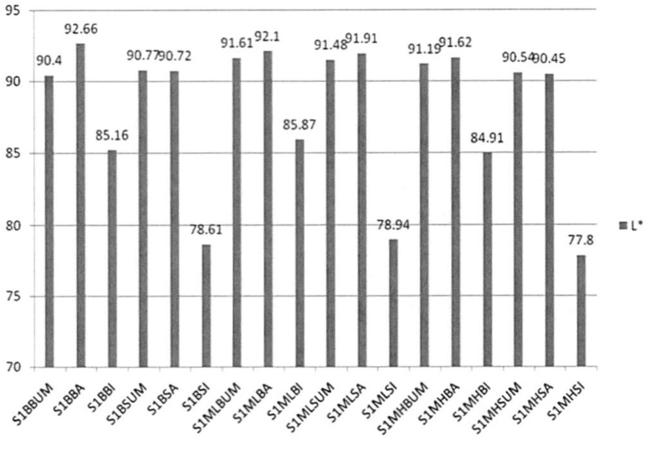

Fig. 4 L* values of S1

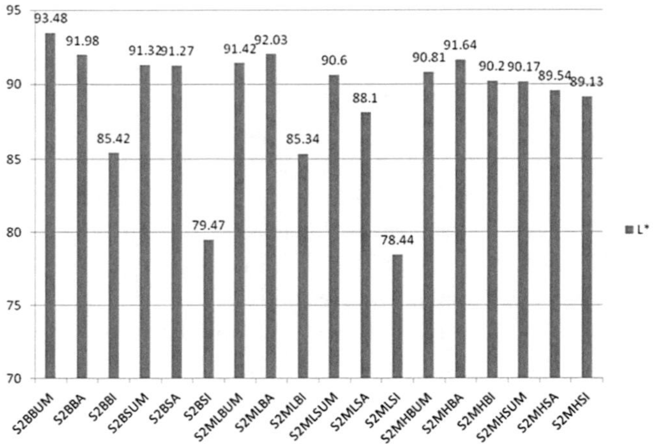

Fig. 5 L* values of S2

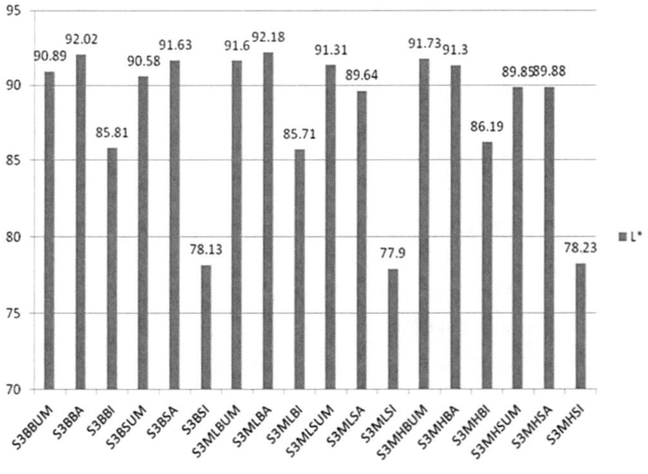

Fig. 6 L* values of S3

darkest shades on bamboo and silk fabrics were obtained from low power (450 W) MAE in which the coordinates of the L* values are 85.34 and 78.44 respectively. Same goes to S3, low power (450 W) MAE showed lower L* values as 85.71 for bamboo fabric and 77.9 for silk fabric.

In general, a* values of the dyed fabrics with natural dye extracted from S1, S2 and S3 are located at positive values (reddish in shade) when the fabrics were treated with iron as mordant. Obviously, most of the untreated fabrics and treated fabrics with alum showed negative values of a* (greenish in shade) as tabulated in Tables 2, 3 and 4. The highest a* value is 3.84 and came from dyed silk with S1 extracted by low power (450 W) MAE, while the lowest a* value of −3.95 came from bamboo fabric dyed with S1 extracted by BWE and treated with alum.

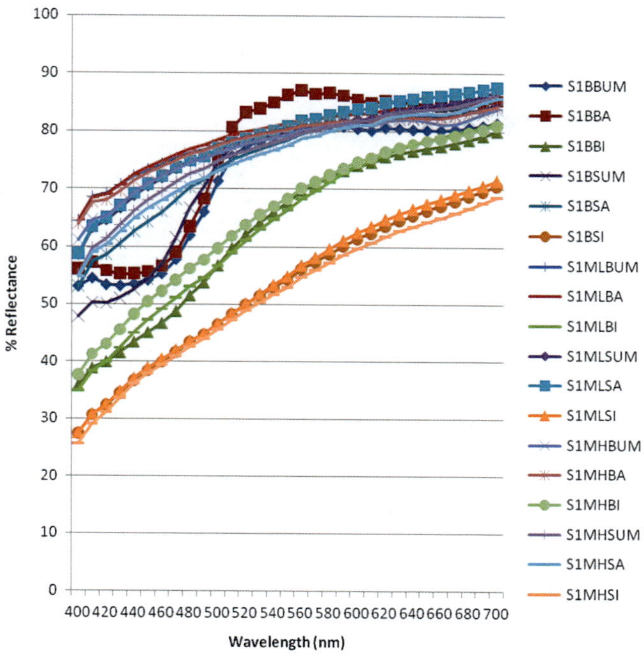

Fig. 7 Reflectance curves of S1

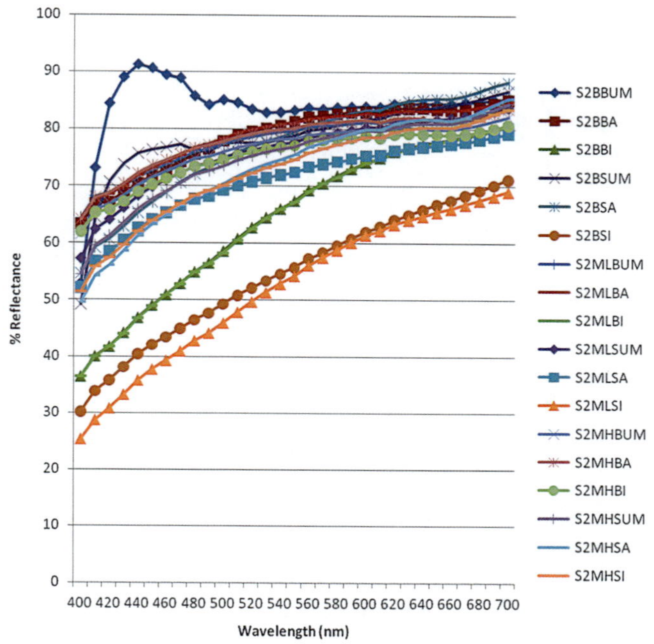

Fig. 8 Reflectance curves of S2

Iron as a mordant also influenced the b* values of the dyed fabrics. Most of the treated fabrics with alum exhibited higher values of b* (positive values which indicate yellowish in shade). However, other dyed fabrics exhibited lower values of b* (positive) except untreated bamboo fabric dyed with natural dye from S2 extracted by BWE.

Reflectance of the Dyed Fabrics

The reflectance curves (% R) for all dyed samples are presented in Figs. 7, 8 and 9. The lower curves indicate that more dyes were absorbed by the fabrics. It is quite obvious that all the dyes showed similar trends because they were from the same source. The lowest % R curve of the dyed fabrics with S1 extract was the silk fabric treated with iron and dyed with high power MAE colourant. Dyed treated silk with iron with S2 extracted by low power MAE showed lowest % R curve. Similar to S3, the lowest % R curve was represented by the treated silk fabric with iron and dyed with low power MAE. There are some agreement with the L* values discussed earlier. Generally, dyes extracted using MAE gave lower % R curve after treating with iron than dye extracted from BWE.

UV–Vis Spectra

The optical absorbance spectrum for S1, S2 and S3 dyes showed the maximum absorption peak in the visible region at 410 and 673 nm wavelength but at different absorbance

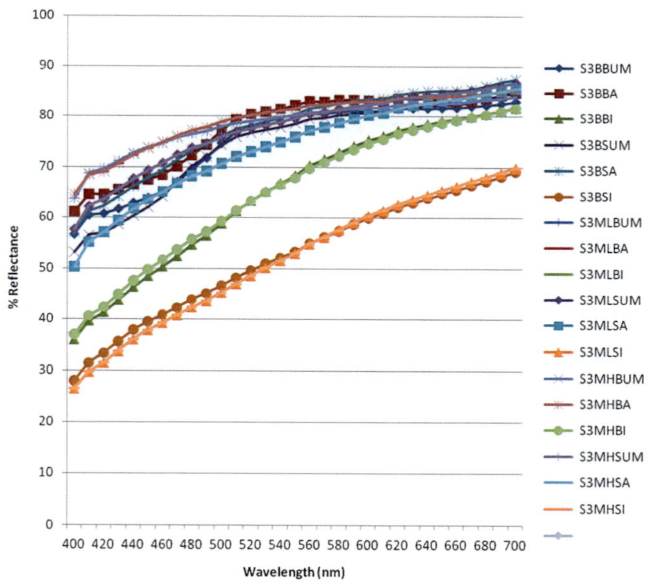

Fig. 9 Reflectance curves of S3

values. Absorbance values for S1 at 410 and 673 nm are 0.35771 and 0.12692, accordingly. S2 showed the lowest absorbance values which are 0.24993 at 410 nm and 0.093677 at 673 nm.

The maximum absorption peaks indicated that the main colourant compound presence in S1, S2 and S3 are carotenoids and chlorophyll a. Carotenoids are represented at 410 nm and chlorophyll a is represented at 673 nm as stated in previous work by Warkoyo and Saati [22] and Das and Govinjee [23] (Fig. 10).

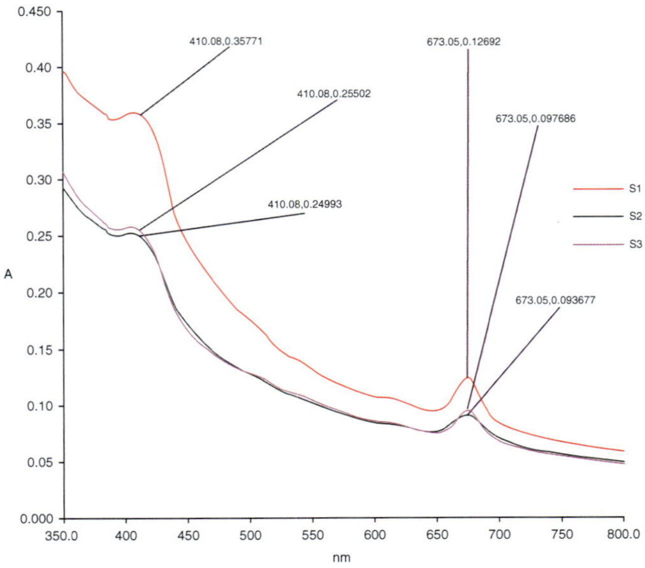

Fig. 10 UV–Vis spectra for S1, S2 and S3

Table 5 Summary of fastness properties to washing of the dyed fabrics

			Washing		
			Colour	Staining	
Extraction	Fabrics	Mordants	change	Cotton	Silk
BWE	Bamboo	No mordant	4/5	4/5	4/5
		Alum	4/5	4/5	4/5
		Iron	4	4/5	4/5
	Silk	No mordant	4/5	5	4/5
		Alum	4/5	4/5	4/5
		Iron	4/5	4/5	4/5
Low power MAE	Bamboo	No mordant	4/5	5	5
		Alum	4/5	4/5	4/5
		Iron	4/5	4/5	4/5
	Silk	No mordant	5	5	5
		Alum	4/5	4/5	4/5
		Iron	4/5	4/5	4/5
High power MAE	Bamboo	No mordant	5	5	5
		Alum	4/5	4/5	4/5
		Iron	4	4/5	4/5
	Silk	No mordant	4/5	5	5
		Alum	4/5	4/5	4/5
		Iron	4	4/5	4/5

Table 6 Summary of fastness properties to perspiration of the dyed fabrics

			Perspiration		
			Colour	Staining	
Extraction	Fabrics	Mordants	change	Cotton	Silk
BWE	Bamboo	No mordant	4/5	4/5	4/5
		Alum	4/5	4/5	4
		Iron	4	4	4
	Silk	No mordant	4/5	4/5	4/5
		Alum	4/5	4/5	4
		Iron	4/5	4/5	4
Low power MAE	Bamboo	No mordant	4/5	4/5	4/5
		Alum	4/5	4/5	4/5
		Iron	4/5	4	4
	Silk	No mordant	4/5	4/5	4/5
		Alum	4/5	4/5	4/5
		Iron	4	4	4
High power MAE	Bamboo	No mordant	4/5	4/5	4/5
		Alum	4/5	4/5	4
		Iron	4	4	4
	Silk	No mordant	4/5	4/5	4/5
		Alum	4/5	4/5	4/5
		Iron	4	4/5	4/5

Table 7 Summary of fastness properties to rubbing/crocking and light of the dyed fabrics

			Rubbing/crocking		
Extraction	Fabrics	Mordants	Dry	Wet	Light
Boiling water	Bamboo	No mordant	4/5	4	3
		Alum	4/5	4/5	3
		Iron	3/4	4	3
	Silk	No mordant	4/5	4	3
		Alum	4/5	4/5	3
		Iron	3/4	4	3
Low power MAE	Bamboo	No mordant	4/5	4/5	4
		Alum	4/5	4/5	4
		Iron	3	3/4	4
	Silk	No mordant	4/5	4	4
		Alum	4/5	4/5	4
		Iron	3/4	4/5	4
High power MAE	Bamboo	No mordant	4	4	4
		Alum	4/5	4/5	4
		Iron	3/4	4	4
	Silk	No mordant	4/5	4	4
		Alum	4/5	4/5	4
		Iron	4	4/5	4

Fastness Properties

The fastness properties of dyed bamboo and silk fabrics are given in Tables 5, 6 and 7. The washing fastness result show that all the dyed samples of S1, S2 and S3 extracts have good to excellent fastness properties and rated between 4 and 5. The fastness properties to perspiration results were also found good to excellent as shown by washing fastness. However, the fastness properties to rubbing/crocking show acceptable results. Obviously, the light fastness of the fabrics is poor which is rated between 3 and 4.

Conclusion

The *K. alvarezii* species can be extracted and utilised as textile colourants on both cellulosic and protein fibres even without mordant. Microwave-assisted extraction produced darker shades in comparison with boiling water extraction. The compound which contributed to the shades obtained is carotenoid and chlorophyll a. Thus, the extract produced greenish yellow and brownish shades on bamboo and silk fabrics. The fastness properties of the dyed fabrics are good to excellent rating except the light fastness properties which is considered poor.

Acknowledgment The authors acknowledged the financial support obtained from the Ministry of Education Malaysia (MOE) under the Exploratory Research Grant Scheme (ERGS). The assistance from the Research Management Institute (RMI) is also highly appreciated.

References

1. A.K. Samanta, P. Agarwal, Application of natural dyes on textiles. Ind. J. Fibre Text. Res. **4**, 384–399 (2009)
2. W.Y. Wan Ahmad, R. Rahim, M.R. Ahmad, M.I. Ab Kadir, M.I. Misnon, The application of Gluta aptera wood (Rengas) as natural dye on silk and cotton fabrics. Univ. J. Environ. Res. Technol. **1**(4), 545–551 (2011)
3. M.I. Ab Kadir, W.Y. Wan Ahmad, M.R. Ahmad, M.I. Misnon, H. Abdul Jabbar, Fastness properties and colorimetric characteristics of low temperature dyeing of natural dyes from the barks of *Ixonanthes icosandra Jack* on polyester fabric, in *2013 IEEE Business Engineering and Industrial Applications Colloquium*, 2013, pp. 430–434
4. R. Shishoo, *Plasma Technology for Textiles* (Woodhead Publishing, Cambridge, 2006)
5. B. Chengaiah, R.K. Mallikarjuna, K.K. Mahesh, M. Alagusundaram, C.C. Madhusudhana, Medicinal importance of natural dyes: a review. Int. J. PharmTech Res. **2**(1), 144–154 (2010)
6. M.D. Telli, R. Paul, P.D. Pardeshi, Natural dyes: classification chemistry and extraction methods. Colourage **48**(4), 51 (2001)
7. R. Bhuyan, C.N. Saikia, K.K. Das, Extraction and identification of colour components from the bark of *Mimusops elengi* and *Terminalia arjuna* and evaluation of their dyeing characteristics on wool. Ind. J. Fibre Text. Res. **29**, 470–476 (2004)
8. R. Balagurunathan, T. Diraviyam, M. Radhakrishnan, Antioxidant activity of melanin pigment from *Streptomyces species* D5 isolated from desert soil, Rajasthan, India. Drug Inven. Today **3**(3), 12–13 (2011)
9. W.A. Ahmad, W.Y. Wan Ahmad, Z.A. Zakaria, Z. Yusof, *Application of Bacterial Pigments As Colorants: The Malaysian Perspective* (Springer, New York, 2012)
10. M. Nerurkar, J. Vaidyanathan, R. Adivarekar, Z.B. Langdana, Use of a natural dye from *Serratia marcescens* subspecies *marcescens* in dyeing of textile fabrics. Octa J. Environ. Res. **1**(2), 129–135 (2013)
11. C. Peter, E.H. Micheal, *Marine Biology*, 8th edn. (McGraw-Hill, New York, 2010), pp. 91–104
12. M.J. Dring, *The Biology of Marine Plants* (Edward Arnold, London, 1982), pp. 1–8
13. S.S. Piyusha, K.R. Vijay, S.G.S. Shelar, M. Kavitha, G.P. Kumar, G.V.S. Reddy, Medicinal value of seaweeds and its applications – a review. Cont. J. Pharmacol. Toxicol. Res. **5**(2), 1–22 (2012)
14. M.D. Guiry, G.M. Guiry, AlgaeBase. World-wide electronic publication, National University of Ireland, Galway. http://www.algaebase.org, Retrieved 03 Feb 2014
15. M. S Doty, The production and use of Eucheuma, in *Case Study of Seven Commercial Seaweeds Resources*, ed. by M.S. Doty, J.F. Caddy, B. Santelices. FAO fisheries technical paper -T281 (Fisheries and Agriculture Department, Rome, 1987), pp. 123–161
16. J. Singh, S. Gu, Commercialization potential of macroalgae for biofuels production. Renew. Sust. Energy Rev. **14**, 2596–2610 (2010)
17. A.D. Hughes, M.S. Kelly, K.D. Black, M.S. Stanley, Biogas from macroalgae: is it time to revisit the idea? Biotechnol. Biofuels **5**, 1–7 (2012)
18. S.M. Arad, A. Yaron, Natural pigments from red microalgae for use in foods and cosmetics. Trends Food Sci. Technol. **3**(4), 92–97 (1992)
19. R. Prasanna, A. Sood, A. Suresh, S. Nayak, B.D. Kaushik, Potentials and applications of algal pigments in biology and industry. Acta Bot. Hung. **49**(1), 131–156 (2007)
20. S. Paoline, J.C. Claire, D. Elie, I. Arsene, Commercial applications of microalgae. J. Biosci. Bioeng. **101**(2), 87–96 (2006)
21. P. Indira, R. Biswajit, Commercial and industrial applications of micro algae – a review. J. Algal Biomass Utiliz. **3**(4), 89–100 (2012)
22. Warkoyo, E.A. Saati, The solvent effectiveness on extraction process of seaweed pigment. Makara Teknologi **13**(1), 5–8 (2001)
23. M. Das, Govinjee, A long-wave absorbing from chlorophyll a responsible for the red drop in fluorescence at 298°K and the F723 band at 77°K. Biochim. Biophys. Acta **143**, 570–576 (1976)

Reducing the Effluent Pollution by Using Trisodium Nitrilotriacetate in Batch Process of Dyeing Cotton Fabric with Fiber-Reactive Dyes

Awais Khatri, Hafeezullah Memon, Zaib-un-Nisa Bhatti, Shakeela Qureshi, and Faisal Zaib

Abstract

A biodegradable organic alkaline salt (trisodium nitrilotriacetate) was used in the batch process of dyeing cotton fabrics with fiber-reactive dyes replacing the conventionally used inorganic salt, sodium chloride, and inorganic alkali, sodium carbonate. The comparative dyeing results were evaluated for color yield, colorfastness, and total dissolved substances (TDS) content of the effluent. Variables affecting dyeing results such as salt/alkali concentration, dye fixation temperature and time were studied. The color yield and colorfastness of the dyed fabrics using trisodium nitrilotriacetate were very similar to those obtained using inorganic salt and inorganic alkali. The reduction in TDS content of the dyeing effluent, when using trisodium nitrilotriacetate, finally supported the aim of the work.

Keywords

Batch dyeing process • Cotton fabric • Fiber-reactive dyes • Polluted effluent • Biodegradable organic salt • TDS content

Introduction

Globally, the textile dyeing industry is known to be a major contributor to environmental pollution [1, 2], mainly because of heavy discharges of inorganic metal salts, bases, acids, other auxiliaries, and organic substances such as dyestuff, to the effluent. Textile dyeing industry also consumes large volumes of clean water [3]. Effluent treatment and water recycling techniques can significantly reduce the pollution and provide water reusable for processing. However, such treatments are costly and generate concentrated solid polluted wastes [4]. Thus, the better way should obviously be to alter the dyeing chemistry and technologies to decrease the effluent pollution.

For apparel-textiles, the predominant dye-fiber combination is that of fiber-reactive dyes and cotton [5]. However, this combination causes the highest discharge per unit-fiber mass of inorganic salts, as electrolyte, and inorganic alkalis, as the dye-fiber reaction activator [6–8]. Such discharge generates environmental problems such as highly colored effluent and heavy contents of the dissolved solids [9, 10]. Various developments have achieved reduction in the need of the amount of inorganic salt with better dye fixation. Popular of them are progress in dye structure development, developments in dyeing processes and machinery [6, 11–16], and cationization of cotton before dyeing [17–21]. The cationization has been proved to be able to eliminate the need of inorganic salt and also inorganic alkali as in certain instances. Some organic substances have also replaced inorganic salts effectively [10, 22–24]. The tetrasodium EDTA, a biodegradable organic alkaline salt, has been studied as an alternative to inorganic salt and inorganic alkali in batch process of dyeing cotton fabric with fiber-reactive dyes [25]. This chapter reports the extension of same approach by using another biodegradable organic alkaline salt, trisodium nitrilotriacetate (trisodium NTA), after successful results by pad-steam process [26]. The optimum dye bath concentrations and dyeing parameters for the organic and inorganic chemicals were determined and compared. The

A. Khatri (✉) • H. Memon • Z.-u.-N. Bhatti • S. Qureshi • F. Zaib
Department of Textile Engineering, Mehran University of Engineering & Technology, Jamshoro 76060, Sindh, Pakistan
e-mail: awais.khatri@faculty.muet.edu.pk

M.R. Ahmad and M.F. Yahya (eds.), *Proceedings of the International Colloquium in Textile Engineering, Fashion, Apparel and Design 2014 (ICTEFAD 2014)*, DOI 10.1007/978-981-287-011-7_19,

total dissolved substances (TDS) content of the effluent was finally tested. As a supplementary experiment, the dyeing results obtained using trisodium NTA with optimum concentration were also compared with those achieved using same concentration of the tetrasodium EDTA.

Materials and Methods

Fabric, Dye, Dye Bath Auxiliaries, and Chemicals

Industry-scoured and -whitened cotton woven fabric (120 g/m^2) was received from Popular Fabrics (Pvt.) Ltd. Pakistan. The fabric was boiled with 2.5 g/l Landipur RSK (nonionic detergent, Clariant Pakistan Ltd.) for 15 min and rinsed with hot then cold water on a Rapid Winch dyeing machine to make it ready-to-dye. CI Reactive Blue 220, a sulfatoethylsulfone-based reactive dye, was obtained from DyStar Pakistan and used without further purification. The Landipur RSK was also used for washing-off after dye application. The commercial grade trisodium NTA (Trilon A 92 R) and tetrasodium EDTA (Trilon B) of BASF were used. The sodium chloride and sodium carbonate were analytical grade.

Batch Dyeing

The fabric samples were dyed by batch dyeing method on a rapid high-temperature-dyeing machine. Dye baths were prepared containing the selected salt and 2 % dye (on mass of fiber, o.m.f.) and 60 g/l sodium chloride with liquor-to-fiber ratio of 15:1. The fabric was immersed and the dye bath was loaded to the rapid HT dyeing machine. The temperature was raised to 40 °C and the dyeing continued for 40 min. Then sodium carbonate was added (2.5 % o.m.f.), the temperature was raised to the dye fixation temperature of 60 °C, and the dyeing was continued for a further 45 min. Various dyeings were performed to study the effect of concentration variation in sodium chloride (0–70 g/l), sodium carbonate (0–30 g/l), and trisodium NTA (0–60 g/l) and the effect of variation in fixation temperature (50–80 °C) and time (30–90 min).

Washing-Off

For removing the unfixed dye and the residual substances, the dyed fabrics were rinsed cold, hot rinsed, soaped (2 g/l Landipur RSK, liquor-to-fiber ratio 60:1, 95 °C, 15 min), and hot rinsed until dye desorption stopped. Finally, the fabric samples were cold rinsed then dried at ambient temperature.

Testing

Color yield (*K/S value*) of the dyed fabric samples was measured on a Gretag Macbeth CE-7000A spectrophotometer. The samples were also tested for colorfastness to crocking (I S O 1 0 5—X 1 2), to washing (I S O 1 0 5—C 0 3), and to light (I S O 1 0 5—B 0 2). The TDS of representative dyeing effluents was tested using a lab TDS meter (Eutech). The representative dyeing effluent was prepared by diluting the optimum dyeing recipe to 100 times.

Discussion on Results

Trisodium NTA is a soluble organic alkaline salt like tetrasodium EDTA which has been reported as an exhausting and fixing agent for batch process of dyeing cotton with fiber-reactive dyes [25]. This is because these sodium salts act as soluble electrolytes and hence substitute sodium chloride or sodium sulfate. Moreover, they provide a basic pH of 10–12 in water; thus, they can activate dye-fiber reaction without adding an inorganic alkali. Accordingly, this study was designed with the consideration of trisodium NTA being exhausting and fixing agent.

Optimum Conventional Dye Bath Formulation

Effect of Sodium Chloride Concentration

Sodium chloride is used as an electrolyte to promote reactive dye exhaustion and improve the ultimate color yield of the cotton dyed fabric [7, 8, 27]. Figure 1 shows the effect of sodium chloride concentration on the color yield of cotton fabric samples dyed with CI Reactive Blue 220 and constant sodium carbonate concentration (20 g/l). The color yield increased up to 60 g/l of sodium chloride after which the increase was negligible.

Effect of Sodium Carbonate Concentration

Sodium carbonate is used as an alkali to activate the dye-fiber reaction in dyeing cotton with fiber-reactive dyes. The absence of alkali has been reported to result very low color yields of the dyed fabric [7, 8, 27–29]. This is because most of the dye molecules on the fiber do not react with cotton polymer and are washed-out during washing-off. Figure 2 shows the effect of sodium carbonate concentration on the color yield of cotton fabric samples dyed with 2 % o.m.f. CI Reactive Blue 220 and constant sodium chloride concentration (60 g/l). The maximum value was obtained at 20 g/l sodium carbonate.

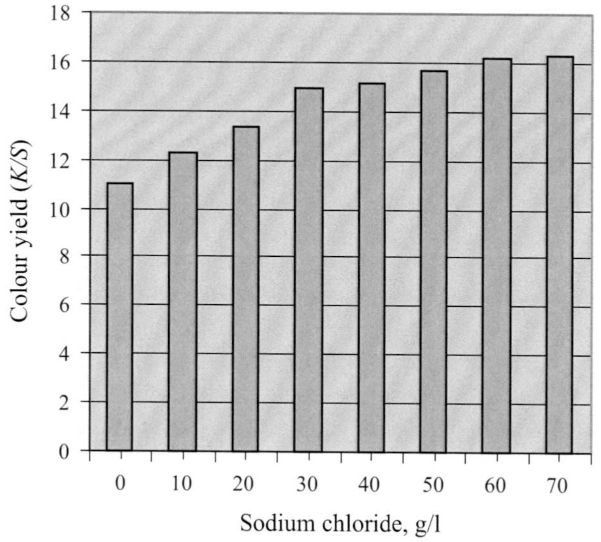

Fig. 1 Effect of concentration of sodium chloride on color yield (2 % o.m.f. CI Reactive Blue 220, 20 g/l sodium carbonate)

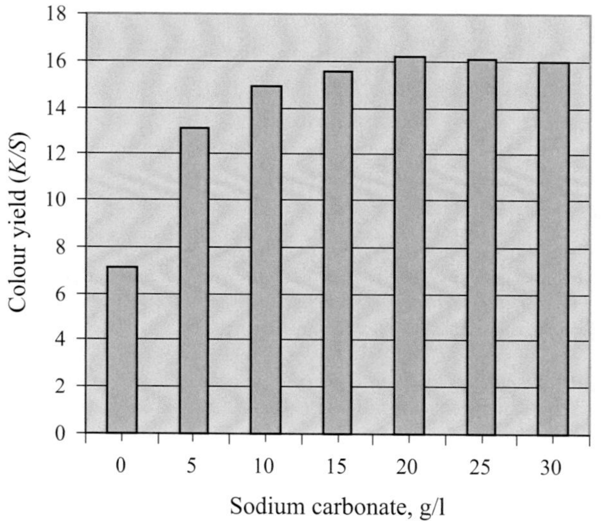

Fig. 2 Effect of sodium carbonate concentration on color yield (2 % o.m.f. CI Reactive Blue 220, 60 g/l sodium chloride)

Effect of Trisodium NTA Concentration

Figure 3 shows that the color yield of cotton fabric dyed with 2 % o.m.f. CI Reactive Blue 220 increased with increasing trisodium NTA concentration and passed through maximum value, which may be due to the result of increasing dye aggregation [30], and competing dye-fiber reaction and dye hydrolysis [7, 8]. The optimum value was obtained with 50 g/l of trisodium NTA. This optimum concentration was lower than the optimum concentrations of the chemicals required for conventional dyeing (i.e., 60 g/l sodium chloride and 20 g/l sodium carbonate).

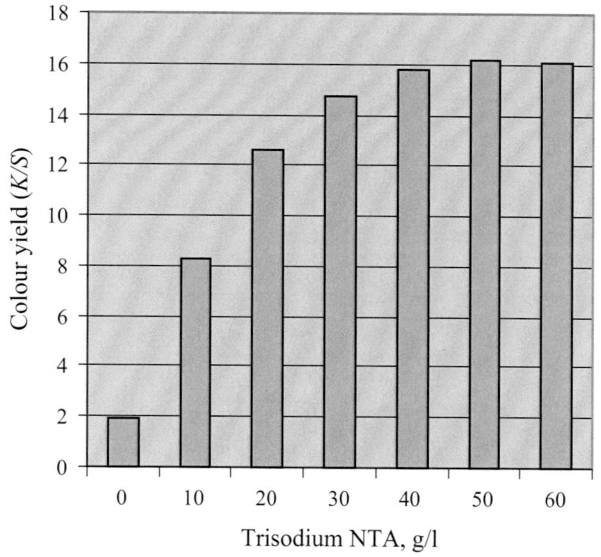

Fig. 3 Effect of concentration of trisodium NTA on color yield (2 % o.m.f. CI Reactive Blue 220)

Optimum Dye Fixation Temperature and Time Using Trisodium NTA

Dye fixation temperature and time are the important parameters in the dyeing of cotton fabric with fiber-reactive dyes. Their optimum values depend on the type of reactive group in the dye structure, mass of the substrate, and liquor-to-fiber ratio of the dye bath. Figure 4a shows the effect of dye fixation temperature on color yield of cotton dyed with 2 % o.m.f. CI Reactive Blue 220 at constant dye fixation time (45 min) and similarly Fig. 4b shows the effect of dye fixation time at constant temperature (60 °C). The optimum temperature was 60 °C, which has been reported as the suitable temperature for sulfatoethylsulfone-based dyes [8]. The color yield continued to increase with increasing dye fixation time. The optimum time was not determined as the industrial dyeings are not performed longer than 90 min of fixation.

Dyeing Using Trisodium NTA Versus Conventional Dyeing

For comparative analysis, the results were obtained by dyeing cotton fabric with 2 % o.m.f. CI Reactive Blue 220 at 60 °C for 45 min using optimum chemical concentrations. The color yield and colorfastness results are given in Table 1. The color yield is almost identical and the colorfastness results are same. Table 2 shows the TDS of dyeing effluents. The results show that trisodium NTA provided around 65 % reduction in effluent TDS. Such reduction in the effluent pollution with no change in the dyeing results is highly encouraging for improving environmental sustainability.

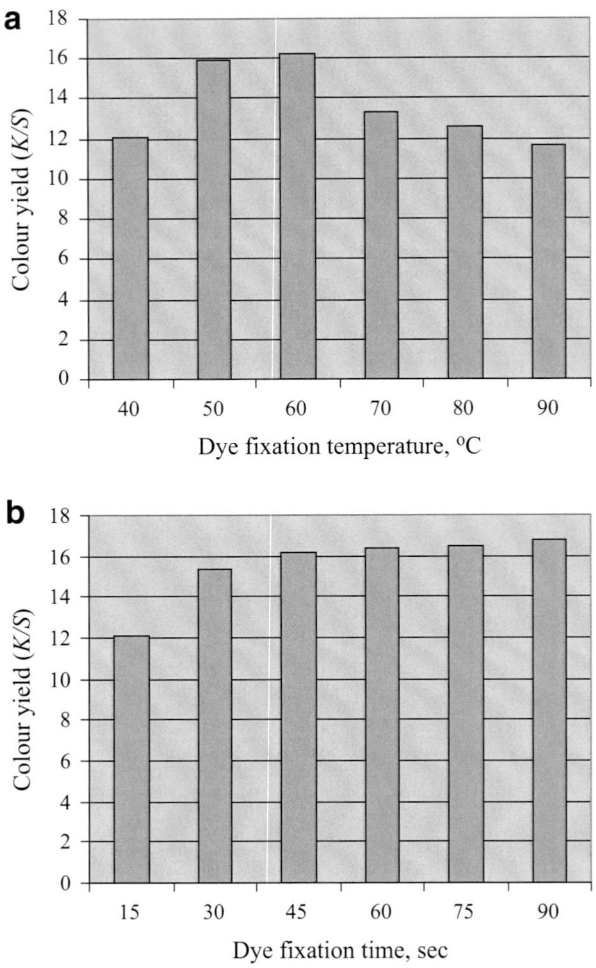

Fig. 4 (**a**) Effect of dye fixation temperature on color yield (2 % o.m.f. CI Reactive Blue 220, 50 g/l trisodium NTA). (**b**) Effect of dye fixation time on color yield (2 % o.m.f. CI Reactive Blue 220, 50 g/l trisodium NTA)

Table 1 Dyeing results using optimum chemical concentration

Process for reactive dyeing of cotton	Color yield (K/S)	Rubbing fastness (grayscale value)		Washing fastness (grayscale value)		Light fastness (blue wool reference value)
		Dry	Wet	Color change	Staining	
Conventional dyeing	16.24	4/5	4	4/5	4/5	7
Dyeing using trisodium NTA	16.20	4/5	4	4/5	4/5	7
Dyeing using tetrasodium EDTA	11.57	–	–	–	–	–

Tables 1 and 2 also show the color yield of the cotton fabric dyed with 2 % o.m.f. CI Reactive Blue 220 and effluent TDS of dye bath using 50 g/l tetrasodium EDTA (previously reported by Ahmed [25]). The TDS reduction was as significant as obtained using trisodium NTA; however, the color yield was not acceptable. This is because Ahmed pretreated the fabric with 5 g/l sodium carbonate (an inorganic alkali) and 3 g/l nonionic detergent for 4 h before dyeing. This severe alkaline treatment possibly reduced the requirement of basic pH for reactive dye fixation in the fiber. Thus, it has been concluded that trisodium NTA is a better biodegradable organic alkaline salt than tetrasodium EDTA for reactive dyeing of cotton fabric.

Table 2 TDS and pH of dyeing effluents

Effluent type	TDS (ppm)	pH
Conventional dyeing	879	8
Dyeing using trisodium NTA	312	8
Dyeing using tetrasodium EDTA	285	8

Conclusions

The trisodium nitrilotriacetate, a biodegradable organic alkaline salt, can be used for batch process of dyeing cotton fabric with fiber-reactive dyes as a substitute to inorganic salt and inorganic alkali. Notable reduction in the TDS content of the (representative) dyeing effluent is a dominant advantage of using trisodium NTA. Interestingly, the trisodium NTA produced even better dyeing results than the tetrasodium EDTA.

References

1. A.D. Broadbent, Water treatment, in *Basic Principles of Textile Coloration*, ed. by A.D. Broadbent, 1st edn. (Society of Dyers and Colourists, Bradford, 2005), pp. 130–151
2. M. Bide, Environmentally responsible dye application, in *Environmental Aspects of Textile Dyeing*, ed. by R.M. Christie, 1st edn. (Woodhead Publishing, Manchester, 2007), pp. 74–92
3. B.R. Babu, A.K. Parande, S. Raghu, T. Prem Kumar, Cotton textile processing: waste generation and effluent treatment. J. Cotton Sci. **11**, 141–153 (2007)
4. M. Bide, Decolorisation of effluent and re-use of spent dyebath, in *Environmental Aspects of Textile Dyeing*, ed. by R.M. Christie, 1st edn. (Woodhead Publishing, Manchester, 2007), pp. 149–190
5. D. King, Dyeing of cotton and cotton products, in *Cotton: Science and Technology*, ed. by S. Gordon, Y.-L. Hsieh, 1st edn. (Woodhead Publishing, Cambridge, 2007), pp. 353–377
6. D. Phillips, Environmentally friendly productive and reliable: priorities for cotton dyes and dyeing processes. J. Soc. Dye. Colour. **112**, 183–186 (1996)
7. J. Shore, Dyeing with reactive dyes, in *Cellulosics Dyeing*, ed. by J. Shore, 1st edn. (Society of Dyers and Colourists, Bradford, 1995), pp. 189–245
8. A.D. Broadbent, Reactive dyes, in *Basic Principles of Textile Coloration*, ed. by A.D. Broadbent, 1st edn. (Society of Dyers and Colourists, Bradford, 2005), pp. 332–357
9. B. Smith, Wastes from textile processing, in *Plastics and the Environment*, ed. by A.L. Andrady, 1st edn. (John Wiley, Hoboken, 2003), pp. 293–295
10. H.G. Prabu, M. Sundrarajan, Effect of the bio-salt trisodium citrate in the dyeing of cotton. Color. Technol. **118**, 131–134 (2002)
11. J.A. Taylor, Recent developments in reactive dyes. Rev. Prog. Color. **30**, 93–107 (2000)

12. Anon, Ciba introduces low-salt reactive dyeing. Text. Indus. Dyeg. South. Afr. **15**, 5–8 (1996)
13. D. Hinks, S.M. Burkinshaw, D.M. Lewis, A.H.M. Renfrew, Cationic fiber reactive dyes for cellulosic fibers. AATCC Rev. **1**, 43–46 (2001)
14. A.H.M. Renfrew, *Reactive Dyes for Textile Fibres*, 1st edn. (Society of Dyers and Colourists, Bradford, 1999)
15. C.B. Anderson, Dyeing reactive dyes using less salt. Am. Dyestuff Rep. **83**, 103–105 (1994)
16. K.V. Wersch, Processes for dyeing cellulosic fibres with reactive dyes. Int. Text. Bull. Dyeing/Printing/Finishing **43**, 21–24 (1997)
17. D.P. Chattopadhyay, R.B. Chavan, J.K. Sharma, Salt-free reactive dyeing of cotton. Int. J. Cloth. Sci. Technol. **19**, 99–108 (2007)
18. S. Zhang, W. Ma, B. Ju, N. Dang, M. Zhang, S. Wu et al., Continuous dyeing of cationised cotton with reactive dyes. Color. Technol. **121**, 183–186 (2005)
19. R.S. Blackburn, S.M. Burkinshaw, Treatment of cellulose with cationic, nucleophilic polymers to enable reactive dyeing at neutral pH without electrolyte addition. J. Appl. Polym. Sci. **89**, 1026–1031 (2003)
20. S.M. Burkinshaw, M. Mignanelli, P.E. Froehling, M.J. Bide, The use of dendrimers to modify the dyeing behaviour of reactive dyes on cotton. Dyes Pigments **47**, 259–267 (2000)
21. M.S.S. Kannan, R. Nithyanandan, Salt and alkali free reactive dyeing on cotton. ATA J. **17**, 60–61 (2006)
22. J.W. Rucker, D.M. Guthrie, Reduction of salt requirements in dyeing cotton with fibre reactive dyes. Presented at the AATCC international conference exhibition, book of papers, 1997
23. J.W. Rucker, D.M. Guthrie, Salt substitute for dyeing cotton with fiber reactive dyes. Sen-I Gakkaishi **53**, 256–260 (1997)
24. Y. Guan, Q.-K. Zheng, Y.-H. Mao, M.-S. Gui, H.-B. Fu, Application of polycarboxylic acid sodium salt in the dyeing of cotton fabric with reactive dyes. J. Appl. Polym. Sci. **105**, 726–732 (2007)
25. N.S.E. Ahmed, The use of sodium edate in the dyeing of cotton with reactive dyes. Dyes Pigments **65**, 221–225 (2005)
26. A. Khatri, R. Padhye, M. White, The use of trisodium nitrilo triacetate in the pad–steam dyeing of cotton with reactive dyes. Color. Technol. **129**, 76–81 (2013)
27. J.R. Aspland, Practical application of reactive dyes. Text. Chem. Color. **24**, 35–40 (1992)
28. D.A.S. Phillips, The dyeing of cellulosic fibres with reactive dyes. Adv. Colour Sci. Technol. **1**, 1–11 (1998)
29. P.J. Dolby, Dyeing of cellulosic fibres with reactive dyes. Text. Chem. Color. **9**, 264–268 (1977)
30. J.D. Hamlin, D.A.S. Phillips, A. Whiting, UV/Visible spectroscopic studies of the effects of common salt and urea upon reactive dye solutions. Dyes Pigments **41**, 137–142 (1999)

Fastness Properties and Color Analysis of Natural Colorants from Actinomycetes Isolates on Silk Fabric

W.F. Wan Yusoff, S.A. Syed Mohamad, and W.Y. Wan Ahmad

Abstract

Actinomycetes are classified as Gram-positive, saprophytic bacteria that can be found distributed in soil, water, and plants. Some actinomycetes are colorful microorganisms that produce a variety of intra and extracellular pigments naturally with different biological functions. These pigments possess biotechnological importance because they can be used as natural colorants as well as an important source for novel antimicrobial agents in place of antibiotics. In this study, two actinomycetes isolated from compost soil and designated as G1A39 and G1A45 were used for extraction of colorants direct from liquid culture (water extracts) and solvent extraction (ethyl acetate and butanol) method. The colorants produced were subjected to dyeing on silk fabrics by using standard procedure for dyeing process. The dyed fabrics were analyzed for their CIE L*a*b* values as well as their colorfastness properties measurement. The result revealed that L* values were high and a*b* values were found within red–yellow zone. Colorfastness to rubbing and perspiration showed excellent ratings between 4 and 5 but poor ratings were observed for light and washing. Color extract from ethyl acetate was proven to produce deep shades on fabrics and further improvement in dyeing process is underway to increase the quality and strength of colorant.

Keywords

Actinomycetes • Natural colorant • Pigments • Dyeing • Colorfastness

Introduction

Natural colorant has acquired an increasing demand all over the world [1]. This is due to the high awareness among industrial player on the negative effects of using synthetic dyes, thus efforts to replace with dyes from natural sources are extensively done. Natural colorant has numerous advantages such as environmental-friendliness, ecofriendliness, and ancient heritage and also has an esthetic approach [2]. Natural colorants possess various advantageous properties such as antimicrobial activity, stability to light, heat, and pH [3, 4]. Likewise, the use of natural dyes as textile material has become highly demanded because of these features. At present, the textile industries produce over 1.3 million tons of dyes and valued at around USD 23 billion, in which almost all were manufactured synthetically [5]. In the past, most natural colorant was predominantly from plant and animal sources. However, fungi, yeast, and microorganism [6–10] can also offer possible alternative sources as natural colorants. Furthermore, microorganism produces more stable pigments such as carotenoids, flavonoids, quinones, and rubramines [11, 12]. Fermentation techniques used to monitor microorganisms growth could be a valuable approach of manufacturing colorants. Since microorganisms produce a large variety of stable pigments, the fermentation techniques often gave higher yields in pigments as well as produced lower residues compared to the use of plants and animals [13]. Moreover, some natural colorants such as anthraquinones have shown

W.F. Wan Yusoff • S.A. Syed Mohamad (✉) • W.Y. Wan Ahmad
Faculty of Applied Sciences, Universiti Teknologi MARA,
Darul Ehsan, Shah Alam, Selangor 40450, Malaysia
e-mail: sharifah459@salam.uitm.edu.my

M.R. Ahmad and M.F. Yahya (eds.), *Proceedings of the International Colloquium in Textile Engineering, Fashion, Apparel and Design 2014 (ICTEFAD 2014)*, DOI 10.1007/978-981-287-011-7_20,

significant antibacterial activity besides providing good color tone and could serve as functional dyes in producing colored antimicrobial textiles [14].

Among microorganisms that can serve as a source of natural colorants are actinomycetes. Actinomycetes are classified as Gram-positive bacteria, aerobic, and the main groups of microorganisms widely distributed throughout the soil [15]. The population and types of actinomycetes are strongly influenced by factors such as soil type, pH, soil temperature, moisture content, organic matter content, ventilation, and cultivation. In addition, Amal et al. [16] noted that actinomycetes were characterized by the production of a variety of colored pigments such as blue, purple, red, yellow, green, brown, rose, and black. The pigment produced can also be dissolved in the medium or can be maintained within actinomycetes cell materials.

An example of pigment from actinomycetes is carotenoids which are produced in a constitutive, light-induced, or cryptic manner found in many terrestrial *Streptomyces* [17]. *Streptomyces* also synthesized and excrete dark pigments melanin or melanoids which considered being useful in characterization of species origin [18–20]. Actinomycetes are also known to produce various types of antibiotics and moreover some of these antibiotics were associated with pigments [21]. Hence, this study focused on the production and the strength of pigments produced by actinomycetes as colorants upon dyeing on silk fabrics.

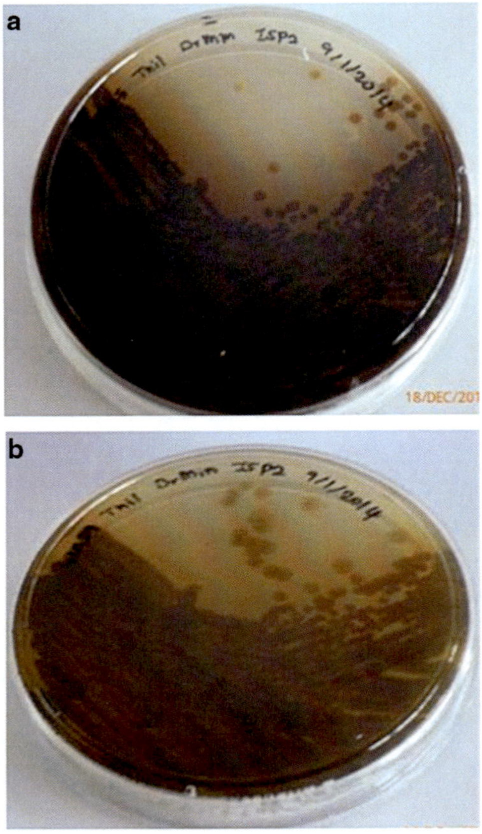

Fig. 1 Growth and mycelium color of two actinomycetes on ISP2 agar (**a**) G1A45 and (**b**) G1A39

Materials and Methods

Materials

Two actinomycetes isolates designated as G1A39 and G1A45 were selected for this study. These two isolates were obtained from goat compost taken from areas around Kuala Selangor, Selangor. Yeast extract, malt extract (ISP2) agar, and starch casein nitrate (SCN) broth medium containing 40 g/L brown sugar were used as media for growth. Ethyl acetate (R & M Chemicals) and butanol (Merck) were used for extraction of the colorant. Fabric dyeing process on silk fabric was carried out using a dyeing machine via the exhaustion method. Four types of colorfastness test were carried out, including washing, light, rubbing, and perspiration following international standards of AATCC. An autowash machine was used to test color fastness to washing, while the lightfastness tester was used to determine the colorfastness of dyed fabrics against sunlight. In addition, crock meter was used to measure the colorfastness to perspiration.

Microorganism Growth

Starch casein nitrate (SCN: 10 g of soluble starch, 2 g KNO_3, 2 g NaCl, 0.2 g casein, 0.05 g of $MgSO_4$ $7H_2O$, 0.02 g $CaCO_3$, 0.01 g of $FeSO_4$ $7H_20$, and 15 g agar in 1 L distilled water) containing 40 g/L stock solution of brown sugar and yeast extract-malt extract (ISP2: 2 g yeast extract, 5 g malt extract, 2 g glucose, and 10 g agar in 1 L distilled water) was used as the growth media. Figure 1 shows the growth of G1A39 and G1A45 on ISP2 agar. A loopful of bacterial cells from single colony of each isolates was inoculated from agar into 300 ml SCN broth containing 40 g/L of brown sugar. The flasks were incubated for 1 week at 30 °C on an orbital shaker at 200 rpm.

Extraction of Pigment

After 1 week of incubation, both mycelia from broth cultures were harvested by centrifugation at 8,000 rpm for 10 min. The mycelium and culture supernatant were both used for colorant

Table 1 Standard method used for colorfastness assessment (Kadir et al. [21])

Colorfastness	Standard Methods	Equipment
Rubbing/crocking	MS ISO 105-X12-2001 MS ISO 105-A04-2003	Crockmeter staining
Perspiration	MS ISO 105-E04-1996 MS ISO 105-A05-2003 MS ISO 105-A04-2003	Perspirometer change in color staining
Washing	MS ISO 105-C01-1966 MS ISO 105-A05-2003 MS ISO 105-A04-2003	Autowash change in color staining
Light	MS ISO-105-B02-2001	Lightfastness tester

extraction. Butanol was added to the mycelium fraction in 1:10 (w/v) ratio and left overnight. Meanwhile, extraction from culture supernatant was carried out by adding ethyl acetate using 1:1 (v/v) ratio and left overnight with shaking at 200 rpm. Afterward, both extracts were collected and concentrated using rotary evaporator (Büchi Rotavapor R-215, Switzerland) at 40 °C with chiller temperature set at below 10 °C. These dried colorants were diluted with distilled water in 1:20 (w/v) ratio before being dyed on silk fabric.

Dyeing and Testing on Fabric

Fabric dyeing was performed using exhaust dyeing techniques on silk fabric. The liquor ratio used was 1:20 to dyebath, while the temperature was set to 80 °C for 1 h. After dyeing, the dyed fabrics were rinsed with tap water and air-dried.

Evaluation and Measurement of Color on Fabric

The dyed fabrics were assessed in accordance to MS ISO standard test methods for colorfastness to perspiration, rubbing/crocking, light, and washing using standard test methods as listed in Table 1 [22]. The color strength (K/S) was measured using HunterLab LabScan XE (LSXE) spectrophotometer and analyzed using HunterLab EasyMatchQC software.

Result and Discussion

Production of Crude Colorants

Extractions of colorant from actinomycetes were carried out using two different solvents which were butanol and ethyl acetate, while water extract from broth medium was used directly as dyes. It is revealed that both solvents produced high concentration of colorant, as compared with water extract. Both isolates produced quite similar orange colorants when extracted by using ethyl acetate solvent. A slightly different color appeared when in butanol where G1A39 produced pink and G1A45 produced wine color. As for water extract, the colors produced by both isolates were dark red. As for G1A45, the weight of colorants obtained was 0.76 g in butanol, while in ethyl acetate only gave 0.59 g. Table 2 shows the concentration of colorant production, crude production as well as volume of colorants produced for dyeing purpose.

The colorants can be produced either extracellularly excreted by the mycelium into the medium or intracellularly within the mycelium. Thus, actinomycetes can produce different type of colorants from different extraction methods. Previous study showed that the colorants obtained from various natural sources like fungi, plants, as well as other species of bacteria were mostly derived from intracellular cells. For instance, a bluish-purple colorant extracted from *Janthinobacteriun lividum* using multiple solvent such as methanol, diethyl ether, acetone, ethanol, and ethyl acetate with different concentration was done by Shirata et al. [6]. A bright red colorant was extracted from *Vibrio* sp. by using methanol and was characterized as prodiginine by Alihosseni et al. [23]. Moreover, Kadir et al. [22] extracted brownish colorant from the bark of *Ixonanthes icosandra Jack* by boiling and methanol extraction.

Dyed Samples

The appearances of colorants from different extracts on dyed silk fabrics are shown in Fig. 2. The extracts gave different shades in which ethyl acetate extract gave the deepest shades which are pale brown for G1A39 and raw umber for G1A45 as compared with other extracts. In butanol, the shades obtained are beaver brown for G1A45 and rosy brown for G1A39. As for water extract, it seem that the shades produced were khaki for G1A45 while G1A39 showing buff in color. All color descriptions were based according to list of shades of brown on Wikipedia website [27].

CIE L*a*b* Value

The CIE L*a*b* values for silk fabrics dyed with colorant from two actinomycetes are listed in Table 3. The L* value indicate perceived lightness or darkness, where 0 indicates black while 100 indicates white. The value of a* indicates red (+a) and green (−a), while b* signifies yellow (+b) and blue (−b). The L* value for water extract of both G1A39 and G1A45 exhibits the highest values which are 72.17 and 71.3, respectively, and this indicates that the colorant is near to white shades. As described earlier, ethyl acetate extract gave the deepest shades; however, they gave high L* values which were 64.14 and 62.16 for G1A39 and G1A45, respectively. The a* and b* values of dyed fabrics presented in Table 3 show that all the samples dyed with actinomycetes extract were found in the red–yellow zone.

Table 2 The extraction and production of colorant from actinomycetes

		Production of colorant			Crude dry weight (g)		Final volume of diluted colorant (g/ml)		
			Solvent extraction		Solvent extraction		Solvent extraction		
No.	Sample	Water extract	Butanol	Ethyl acetate	Butanol	Ethyl acetate	Butanol	Ethyl acetate	Water extract
1.	G1A39	++ (Dark red)	+++ (Pink)	+++ (Orange)	0.76	0.59	15.2	11.8	50
2.	G1A45	++ (Dark Red)	+++ (Wine)	+++ (Orange)	1.34	1.44	26.8	28.8	50

Rating for colorant concentration: + = low, ++ = moderate, +++ = high

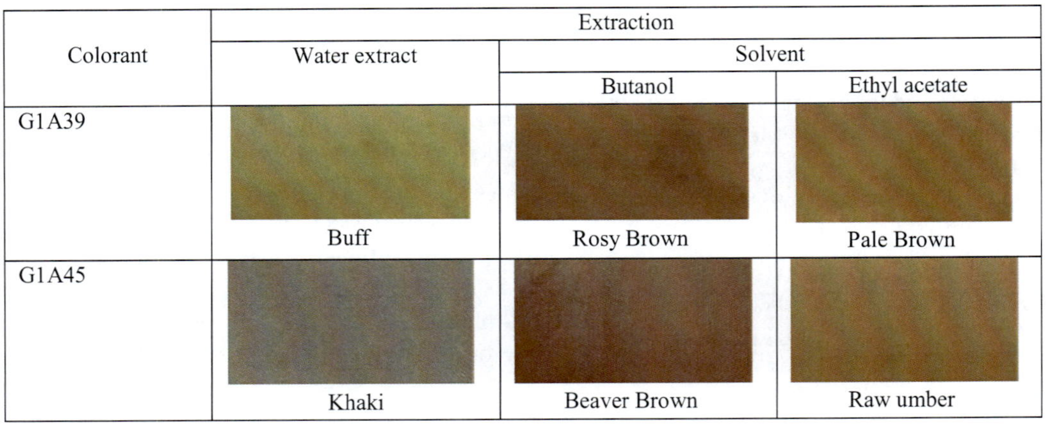

Fig. 2 Silk samples dyed with colorant extracted from actinomycetes

Table 3 L*a*b* value for dyed silk fabric with actinomycetes extract colorant

Samples	Solvent	L*	a*	b*
G1A39	Ethyl acetate	64.14	11.2	24.1
	Butanol	61.62	11.93	16.95
	Water extract	72.17	2.4	29.49
G1A45	Ethyl acetate	62.16	12.47	25.45
	Butanol	64.52	12.6	11.96
	Water extract	71.3	7.19	11.92

Table 4 Summary of fastness properties to rubbing/crocking and light of the dyed silk fabric

		Rubbing/crocking		
Samples	Solvent	Dry	Wet	Light
G1A39	Ethyl acetate	5	5	3/4
	Butanol	4	4	2/3
	Water extract	5	5	2/3
G1A45	Ethyl acetate	5	5	2/3
	Butanol	5	5	2
	Water extract	5	4	2

As a comparison, colorant extracted from onion, Canadian goldenrod, and pomegranate as studied by Mahmud-Ali et al. [24] produced the low a* value which was in the (−a) range of green zone as well as lower L* value which indicates darker shades. Similar results were obtained of *Mahonia napaulensis* stem cells showing the (−a) and (+b) values when dyed on silk and cotton fabrics indicating green–yellow zone although when these colorants were treated with various mordants, the a* value increased to (+a) which within the red zone. However L* value for these colorant were more than 50 indicating whiter shades [25].

Colorfastness Properties

The dyed fabrics were evaluated for colorfastness to rubbing, washing, perspiration, and light. The evaluations were based on the MS ISO standards with rating range between 1 and 5, where 1 is the poorest and 5 is the best. The results for rubbing and lightfastness assessment are listed in Table 4. The rating for fastness to wet and dry rubbing was between 4 and 5 which indicates good to excellent result. For colorfastness to light, the fabrics were assessed by comparing with blue wool standards with rating ranges of 1–8 in which 1 is the lowest and 8 is the highest. The results showed that the dyed silk in general exhibits poor lightfastness giving ratings between 2 and 3/4. Generally, it is well known that the lightfastness properties were poor among natural dyes which usually gave low rating. For instance, bluish-purple colorant extracted from *Janthinobacterium lividum* showed the rating as low as 1 [5], while colorant from the barks of *Ixonanthes icosandra* Jack gave a rating of 2 for lightfastness properties [21]. Conversely, the lightfastness of dyes from lichen gave a good rating from 6 to 7 [26].

Table 5 Summary of fastness properties to perspiration of the dyed silk fabric

		Perspiration		
			Staining	
Samples	Solvent	Color change	Silk	Cotton
G1A39	Ethyl acetate	5	5	4
	Butanol	5	5	4/5
	Water extract	5	5	4/5
G1A45	Ethyl acetate	5	5	4
	Butanol	4/5	5	4
	Water extract	5	5	4

Table 6 Summary of fastness properties to washing of the dyed silk fabric

		Washing		
			Staining	
Samples	Solvent	Color change	Silk	Cotton
G1A39	Ethyl acetate	3	5	4/5
	Butanol	1/2	5	5
	Water extract	3	5	5
G1A45	Ethyl acetate	2/3	5	5
	Butanol	2/3	5	5
	Water extract	2/3	5	5

Table 5 shows the summary of colorfastness to perspiration. The result for fastness to perspiration gave good to excellent rating of 4/5–5 for change in color, while the rating for staining ranged from 4 to 5. Comparatively, bluish-purple colorant extracted from *Janthinobacterium lividum* gave a moderate rating of perspiration fastness properties [5]. The result for colorfastness to washing gave a rating in between 1/2 and 3 for change in color. This significantly indicates that the rating is relatively poor to moderate. However, for staining, the rating for both silk and cotton was high, from 4/5 to 5. From the result, it can be suggested that the colorant faded easily in soapy water, but it did not cause stains in silk and cotton. However, Zulkifli et al. [10] showed a good rating of 4 for red and purple colorants extracted from *Serratia marcescens* and *Chromobacterium violaceum*. As a summary, although the ratings for colorfastness to light and washing were low, the rating against perspiration and rubbing was excellent giving ratings in between 4 and 5 for both colorfastness assessments. Nevertheless, further improvement will be done to strengthen the quality of the dyes, especially for colorfastness to light and washing (Table 6).

Conclusion

From the study, it can be concluded all the fabrics dyed with colorants extracted from actinomycetes produced good colorfastness properties except for light and washing. Ethyl acetate was found to give the best shades on the fabrics, while water extract gave the lightest. Although colorfastness to light and washing gave low rating, the ratings for fastness against rubbing and perspiration were excellent. As a suggestion, modification will be carried out, for instance, treating the colorant with mordant to improve the quality of colorant as well as to use new and cheap extraction method for colorant production.

Acknowledgment The authors would like to acknowledge the Research Management Institute (RMI) of Universiti Teknologi MARA for the financial support under the Excellence Fund Scheme.

References

1. C. Verma, V. Venkatachalam, Effect on mordants on mango (bark) dye for dyeing of jute-cotton union fabric. Colourage **49**, 49–54 (2002)
2. A. Mukherjee, Opportunities of natural dyes for a greener textile and colouration industry. Eco-N-Viron, Serampore, Hooghly. National Workshop and Seminar on Vegetable Dye and its Application on Textiles **70**, 70–73 (2011)
3. P. Murugkar, Z.P. Bhathena, N. Kanongo, R Adivarekar, Isolation of a colour producing microbe for dyeing textiles. J. Textile Assoc. 29–32 (2006)
4. A. Pandey, S. Babitha, Microbial pigments. Adv. Biotechnol. **17** (2005)
5. C.K. Venil, Z.A. Zakaria, W.A. Ahmad, Bacterial pigments and their applications. Process Biochem. **48**, 1065–1079 (2013)
6. A. Shirata, T. Tsukamoto, H. Yasui, T. Hata, S. Hayasaka, A. Kojima, H. Kato, Isolation of bacteria producing bluish-purple pigment and use for dyeing. Jpn. Agric. Res. Q. **34**(2), 131–140 (2000)
7. D. Rettori, N. Durán, Production, extraction and purification of violacein: an antibiotic pigment produced by *Chromobacterium violaceum*. World J. Microbiol. Biotechnol. **14**, 685–688 (1998)
8. F.A. Nagia, R.S.R. El-Mohamedy, Dyeing of wool with natural Anthraquinone dyes from *Fusarium oxysporum*. Dyes Pigments **75**, 550–555 (2007)
9. M. Moss, Bacterial pigments. Microbiologist, 10–12 (2002)
10. R.P. Williams, J.A. Green, D.A. Rappoport, Studies on pigmentation of *Serratia marcescens*. 1. Spectral and paper chromatographic properties of prodigiosin. J. Bacteriol. **71**(1), 115–120 (1956)
11. N.A. Zulkifli1, W.Y.W. Ahmad, W.A. Ahmad, Z.A. Zakaria, J. Salleh, M.R. Ahmad, M.I.A. Kadir, R.A. Halim, Natural colorants from microorganism. *Regional Academic Conference* (2009)
12. N. Durán, M.F.S. Teixeira, R. Conti, E. Esposito, Ecological-friendly pigments from fungi. Crit. Rev. Food Sci. Nutr. **42**(1), 53–66 (2002)
13. D.K. Hobson, D.S. Wales, Green colorants. J. Soc. Dyers Colour **114**, 42–44 (1998)
14. R.J.N. Frandsen, N.J. Nielsen, N. Maolanon, J.C. Sorensen, S. Olsson, J. Nielsen et al., The biosynthetic pathway for aurofusarin in *Fusarium graminearum* reveals a close link between the naphthoquinones and naphthopyrones. Mol. Microbiol. **61**, 1069–1080 (2006)
15. M. Arifuzzaman, M.R. Khatun, H. Rahman, Isolation and screening of actinomycetes from Sundarbans soil for antibacterial activity. Afr. J. Biotechnol. **9**(29), 4615–4619 (2010)

16. A.M. Amal, K.A. Abeer, H.M. Samia, A.E.H. Nadia, K.A. Ahmed, H.M. El-Hennawi, Selection of pigment (Melanin) production in Streptomyces and their application in printing and dyeing of wool fabrics. Res. J. Chem. Sci. **1**(5), 22–28 (2011)
17. S. Dharmaraj, B. Ashokkumar, K. Dhevendaran, Fermentative production of carotenoids from marine actinomycetes. Iran. J. Microbiol. **4**(1), 36–41 (2009)
18. G.M. Zonova, Melanoid pigments of actinomycetes. Mikrobiologiya **34**, 278–283 (1965)
19. T. Aria, Y. Mikami, Choromogenecity of *Streptomyces*. Appl. Microbiol. **23**, 402–406 (1972)
20. S.G. Dastager, L. Wen-Jun, A. Dayanand, T. Shu-Kun, T. Xin-Peng, Z. Xiao-yang, X. Li-Hua, J. Cheng-Lin, Separation, identification and analysis of pigment (melanin) production on *Streptomyces*. Afr. J. Biotechnol. **5**(8), 1134–1134 (2006)
21. J. Miyaura, C. TatsumI, Studies on the antibiotics from actinomycetes An antibiotics pigment from *Streptomyces* F-23b. Bull. Univ. Osaka Pref. Ser. B **12**, 129–137 (1960)
22. M.I.A. Kadir, W.Y.W. Ahmad, M.R. Ahmad, M.I. Misnon, H.A. Jabbar, Fastness properties and colorimetric characteristic of low temperature dyeing of natural dyes from the barks of *Ixonanthes icosandra Jack* on polyester fabric. *IEEE Business Engineering and Industrial Applications Colloquium (BEIAC)*, pp. 549–563 (2013)
23. F. Alihosseini, K.S. Ju, J. Lango, B.D. Hammock, G. Sun, Antibacterial colorants: characterization of prodiginines and their applications on textile materials. Biotechnol. Prog. **24**(3), 742–747 (2008)
24. A.M. Ali, C.F. Binder, T. Bechtold, Aluminium based dye lakes from plant extracts for textile coloration. Dyes Pigments **94**, 533–540 (2012)
25. P.S. Vankar, R. Shanker, S. Dixit, D. Mahanta, S.C. Tiwari, Sonicator dyeing of modified cotton, wool and silk with *Mahonia napaulensis DC*. And identification of the colorant in Mahonia. Ind. Crop. Prod. **27**, 371–379 (2008)
26. N.A. Mohamed, W.Y.W. Ahmad, K. Ngalib, M.R. Ahmad, M.I.A. Kadir, A. Ismail, Dyeing of silk fabric with natural dyes from Lichen using ultrasound assisted extraction (unpublished)
27. http://en.wikipedia.org/wiki/Brown

Dyeing of Polyester and Polyester Microfibre with Natural Dye from Bacteria Source

W.Y. Wan Ahmad, M.R. Ahmad, and M.I. Ab Kadir

Abstract

It is difficult to dye polyester with natural dyes and even more difficult to dye at temperature of 100 °C or above due to the hydrophobicity nature of polyester. In this study, two dyes produced from bacteria were used to dye polyester and polyester microfibre at boiling point without using mordant. The *Chromobacterium violaceum* and *Serratia marcescens* produced purple and red/pink colours, respectively, when applied on polyester and polyester microfibre in crude/paste and liquid forms. The colours produced were analysed using colour and uv-visible spectrophotometers. The colourfastness tests showed that washing and crocking fastness were good but lightfastness was poor. It was found out that *Serratia marcescens* can be used as dye in liquid and paste forms, but *Chromobacterium violaceum* can only be used in crude form.

Keywords

Natural dyes • Bacteria • Dyeing • Polyester • Colourfastness

Introduction

One of the considerations for a beautiful textile product is the coloration of the material which comes in the form of colours and shades. Colours on textiles come from dyes and/or pigment. Natural fibres such as cotton and silk are much easier to dye in comparison with synthetics such as polyester due to the chemical nature of natural fibres. Man-made or synthetic fibres need special dye to colour them because of their properties such as hydrophobic, non-absorbance and the absence of dye sites. Disperse dye which is synthetic is used to dye polyester. The drawbacks of synthetic dyes are mostly based on environmental issue which makes way for natural dyes to make a comeback to counteract the issue of toxicity and effluent problems caused by dyeing with synthetic dyes. Natural dyes come from plants, animals and mineral sources. Natural dyes are best applied to protein fibres such as silk and wool and with a little modification can be used on cotton and other cellulosic fibres but not on synthetic fibres such as polyester [1–3]. Few researchers worked on dyeing of synthetics with natural dyes [4–6] with limited success due to the difficulty in dyeing of synthetic fibres. However, two dyes from bacteria which can be categorised as natural dye from animal were chosen to dye polyester and polyester microfibre. The two bacteria strains are known as *Chromobacterium violaceum* and *Serratia marcescens* which produced purple and red/pink colours, respectively. *Chromobacterium violaceum* was obtained from waste water, and *Serratia marcescens* was obtained from soil [7–9]. The structures of the dyes are depicted in Figs. 1 and 2.

Research on colours from bacteria has been conducted by several workers. Vaidyanathan et al. [12] investigated colours from *Serratia marcescens* which produced red prodigiosin pigments, whilst Shirata et al. [10] and Adivarekar et al. [13] studied on violet pigments produced

Contract grant sponsor: Ministry of Education (MOE) under Exploratory Research Grant Scheme (ERGS), contract grant number 600RMI/ERGS5/3/(11/2011)

W.Y. Wan Ahmad (✉) • M.R. Ahmad • M.I. Ab Kadir
Faculty of Applied Sciences, Textile Research Center, Universiti Teknologi MARA, 40450 Shah Alam, Selangor Darul Ehsan, Malaysia
e-mail: wanyunus@salam.uitm.edu.my

M.R. Ahmad and M.F. Yahya (eds.), *Proceedings of the International Colloquium in Textile Engineering, Fashion, Apparel and Design 2014 (ICTEFAD 2014)*, DOI 10.1007/978-981-287-011-7_21,

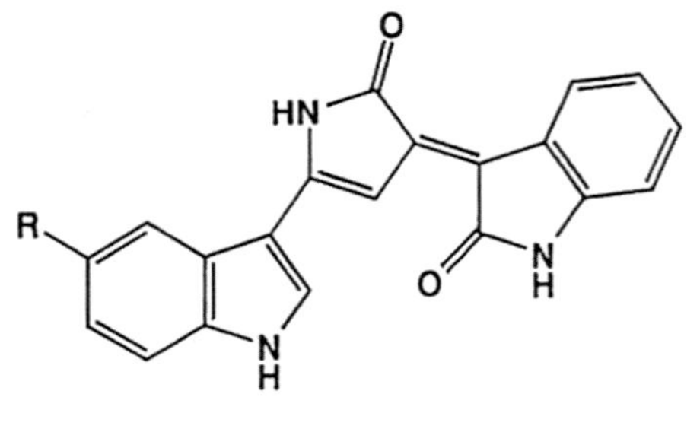

Fig. 1 The coloured component of *Chromobacterium violacein* [10]

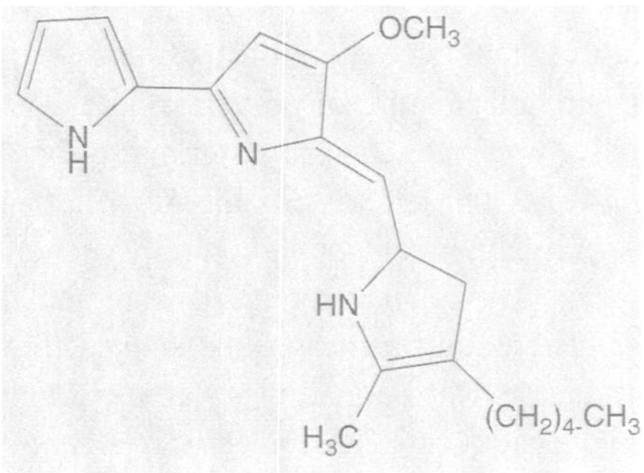

Fig. 2 The coloured component of *Serratia marcescens* [11]

by *Janthinobacterium lividum* and *Chromobacterium* species, respectively. Diana et al. [14], on the other hand, discovered yellow riboflavin from *Ashbya*, and Adivarekar et al. [15] found bacteria with anthraquinones structure from *Fusarium oxysporum*. Nerurkar et al. [16] discovered red rose colour from subspecies of *Serratia marcescens*, and in Malaysia Venil et al. [17] described the use of bacteria not only in textiles but also in food and pharmaceutical, among others. One thing to take note is since most of the researchers were not from the textile coloration field, they used the term dye and pigment interchangeably to refer to colourant they just produced. In textiles, the simple way to differentiate between dyes and pigments is that dyes are soluble in water and do not need binders to attach their colour to substrates whereas pigments are insoluble and do need binders. Furthermore the size of pigment molecules is bigger than dye molecules.

Materials and Methods

Chromobacterium violaceum with purple colour of the violacein type was obtained from waste water. *Serratia marcescens* bacteria strain with red/pink colour of prodigiosin type was obtained from soil. Lightweight polyester and polyester microfibre fabrics for shirting were used in the exhaustion dyeing process. Polyester microfibre is normally more difficult to dye.

Microbial Production

Nutrient agar (NA) was used to prepare active cultures of both bacteria strains. A loopful of bacteria cells from a single colony for each strain was inoculated from NA plate into 25 ml nutrient broth (NB) medium in a 250 ml Erlenmeyer flask separately. The time taken for growth was 12 h at room temperature for *Serratia marcescens* and 30 °C for *Chromobacterium violaceum* on an orbital shaker at 200 revolutions per minute (rpm).

After 12 h of incubation, 10 ml active culture of *Serratia marcescens* was aseptically transferred to a 1,000 ml Erlenmeyer flask which contains 9 ml brown sugar (BS) with concentration of 40 g/L and 81 mL of sterile distilled water. The flask was then incubated at room temperature for another 24 h on an orbital shaker at 200 rpm to grow the bacteria. Brown sugar was used as the carbon source instead of NB because the bacteria produced redder dye in brown sugar medium than the NB medium.

The same methodology was applied to *Chromobacterium violaceum* where 10 ml active culture of the bacteria was also aseptically transferred to a 1,000 mL Erlenmeyer flask containing 90 mL of NB. Then, the incubation of the bacteria took place by shaking on an orbital shaker also at 200 rpm for 24 h at 37 °C [7, 8].

Dye Extraction

The bacteria were harvested after 24 h of incubation. They were centrifuged separately at 7,000 rpm and 4 °C temperature. The extraction processes produced pellets, and methanol was added to the pellets where the mixture was centrifuged also at 7,000 rpm and at 4 °C temperature. The extraction process was repeated until the pellet turns white. This method of extraction was applied to both *Chromobacterium violaceum* and *Serratia marcescens* to produce colours of purple and red/pink dyes. Rotary evaporator was used to remove methanol from dyes and produced the dyes in the form of crude or paste. The time taken for evaporation was 1 h for 250 mL at 45 °C. The production of crude for 1 L of solution was approximately 0.2 g [8].

Dyeing of Fabric Samples

Dyeing at boiling point (100 °C) was carried out for all fabric samples with additional temperature of 130 °C for liquid *Serratia marcescens* dye on both polyester and polyester microfibre. The shade of the sample produced was based on the weight of fabric (owf), and the liquor ratio was 1:20. One hour was the time taken for the dyeing cycle. No addition of mordant or other chemical auxiliaries was used in the dyeing. The *Serratia marcescens* dye can be used in liquid and crude/paste forms to dye both polyester and polyester microfibre, but *Chromobacterium violaceum* dye can only be used to dye samples in crude/paste form [8].

Fastness properties of the dyed fabric were tested according to MS ISO standard. The tests that were carried out were colourfastness to washing, perspiration, light and rubbing (crocking). Colourfastness tests were carried out according to MS ISO 105-C01-1996 for washing, MS ISO 105-E04-1996 for perspiration, MS ISO 105-B02-2001 for light and MS ISO 105-X12-2001 for rubbing or crocking.

Spectrophotometers

The appearance of *Chromobacterium violaceum* and *Serratia marcescens* in methanol was purple and red/pink. They were subjected to PerkinElmer Lambda25 1.27 UV-visible spectrophotometer to determine the wavelength of maximum absorbance from the absorption spectra and to determine the characteristic value of the absorptivity of a specific molecule.

In order to assess the colour, the CIE L*a*b* was used to obtain reflectance curves of the dyed polyester and polyester microfibre samples using GretagMacbeth Color-Eye 7000A.

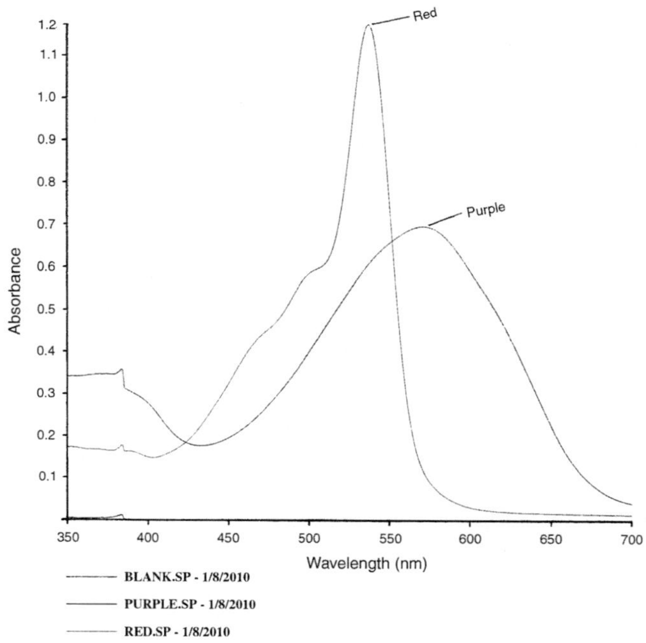

Fig. 3 Absorption spectra of *Chromobacterium violaceum* and *Serratia marcescens* in methanol

Table 1 CIE L*a*b* values for *Chromobacterium violaceum* and *Serratia marcescens*

Dyes and substrates	L*	a*	b*
[a]Red crude polyester 100 °C	67.817	36.099	−22. 071
[b]Purple crude polyester 100 °C	61.306	11.479	−31.136
[a]Red crude microfibre	74.975	35.497	−21.809
[b]Purple crude microfibre 100 °C	81.778	6.218	−22.694
[a]Red liquid polyester 100 °C	81.048	20.114	−14.852
[a]Red liquid microfibre 100 °C	88.540	16.794	−9.770
[a]Red liquid polyester 130 °C	79.061	28.617	−14.527
[a]Red liquid microfibre 130 °C	87.605	20.912	−10.683

[a]Red is *Serratia marcescens*
[b]Purple is *Chromobacterium violaceum*

Results and Discussion

UV-Visible Spectrophotometers

Figure 3 shows broad absorption spectrum ranging from 500 to 700 nm. The maximum wavelength for *Chromobacterium violaceum* in Fig. 3 is 575 nm. Investigation by Lu [18] showed that *Chromobacterium violaceum* colourant in methanol produced maximum absorbance at 576 nm (Lu 9). *Serratia marcescens* maximum wavelength absorbance was at 540 nm [19], while in Fig. 3 the wavelength is at 534 nm. Therefore, both maximum absorbance wavelengths are in agreement with previous findings by other researchers to indicate that bacteria colourants are from the same species.

Colour Measurement

The L* values indicate perceived lightness or darkness. The value of 0 represents black and 100 represents white. The value of (+a*) is red and (−a*) is green, while (+b*) is yellow and (−b*) is blue. Table 1 shows the L*a*b* of both bacteria dyes on polyester and polyester microfibre.

The reading of L* (Table 1) shows that the paste form of *Serratia marcescens* has lower L* values than the liquid form of *Serratia marcescens*. In terms of substrates, polyester sample has lower L* values than polyester microfibre. Both of these L* values relate to the % reflectance in Fig. 4. The lower the % of reflectance, the deeper the shade on the fabric sample. In this case the shade from crude/paste is darker than from liquid, and colour strength on polyester is higher than on polyester microfibre. The higher temperature of dyeing (130 °C) produced lower L* on both polyester and polyester microfibre.

This is related to the absorption of the dye on substrates where darker/deeper shades or lower % reflectance indicates that more dyes absorbed onto the samples. Thus bacteria

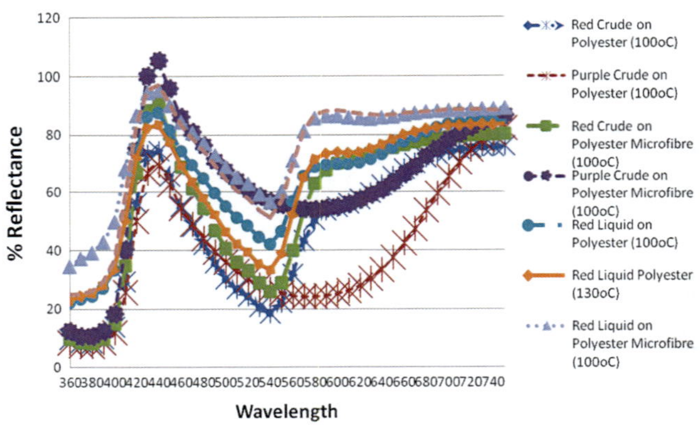

Fig. 4 Percent reflectance of bacteria dyes of *Serratia marcescens* and *Chromobacterium violaceum* on polyester and polyester microfibre

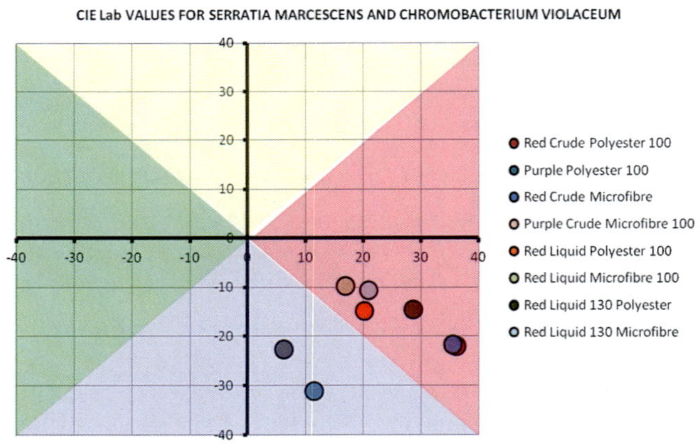

Fig. 5 The 2-D coordinate of *Chromobacterium violaceum* and *Serratia marcescens* on polyester and polyester microfibre

dyes in crude/paste form were preferably absorbed by substrates than in liquid form. On the other hand, polyester substrate accepted more dyes than polyester microfibre counterpart. This is true since polyester microfibre is difficult to dye due to smaller fibre sizes and, in synthetic dyes, a different derivative of disperse dyes was manufactured for microfibre.

The 2-D diagram plotted in Fig. 5 further shows the colour of the dyed samples. Two of the *Chromobacterium* are on the blue 'triangle', while 6 of the *Serratia marcescens* are on the red 'triangle'. The *a and –b* triangles indicate that the samples are of bluish and reddish colours. Figure 6 shows the coloured samples produced from dyeing with both bacteria dyes.

Fig. 6 Polyester samples on cabbage patch kids dyed with *Serratia marcescens* and *Chromobacterium violaceum*

Colourfastness

Overall, the colourfastness to washing was good ranging from 4 to 4/5 for all types of fabrics including polyester and polyester microfibres. Colourfastness to perspiration and colourfastness to crocking were in the range of 4/5. However colourfastness to light was low. The range was below 4 where the highest fastness rating on polyester was 4. Shirata et al. [10] had the same problem on lightfastness indicating a rating of 1 for the purple bacteria. Work to improve lightfastness has been documented in Japanese Patent (JP, 11-152687, A). This is one aspect one has to look upon for future research in dealing with bacteria colourant.

Conclusion

Natural dyes from bacteria can be used to dye polyester and polyester microfibre at boiling point without any auxiliary which is not possible even for synthetic dye. The paste form gave better colour strength in comparison with the liquid form of bacteria dyes of the same species on the same substrate. The bacteria dyes dyed polyester deeper than polyester microfibre with unusually bright colour for natural fibre. One good aspect of bacteria dyes is that not many natural dyes come in reddish and bluish colour and as bright as the colour from bacteria. However, lightfastness was poor. The wash fastness, perspiration fastness and crocking were good. This bacteria dye will not pollute the environment because no mordant or other auxiliaries/assistants were used as in other natural dyes.

Acknowledgement The authors would like to thank the Ministry of Education, Malaysia (MOE), for the research fund. The assistance from the Research Management Institute (RMI), Universiti Teknologi MARA (UiTM) is highly acknowledged.

References

1. S. Cage, Dyeing of wool and other natural fibres with natural dyes, Birmingham guild of weavers spinners and dyers, http://www.creative-chemistry.org.uk/activities/documents/naturaldyeing.pdf. Retrieved 155 May 2012
2. The Maiwa Guide to Natural Dyes, Natural dyes – fibres (2010), http://maiwahandprints.blogspot.com/2010/11/natural-dyes-fibres.html. Retrieved 15 May 2012
3. Maiwa Handprints, Natural dyes what they are and how to use them (n.d.), http://www.maiwa.com/pdf/natural_dyeing.pdf. Retrieved 22 May 2012
4. H.T. Lokhande, V.A. Dorugade, S.R. Naik, Application of natural dyes on polyester. Am. Dyestuff Rep. **87**(9), 40–48 (1998)
5. D.R. Rathi, R.N. Padhye, Studies in application of natural dyes on polyester. Colourage **41**, 25–28 (1994)
6. H.T. Lokhande, V.A. Dorugade, Dyeing nylon with natural dyes. Am. Dyestuff Rep. **88**(2) 1999
7. N.A. Zulkifli, W.Y. Wan Ahmad, W.A. Ahmad, Z.A. Zakaria, J. Salleh, M.R. Ahmad, M.I. Ab Kadir, R. Abdul Halim, Natural colorants from microorganism. Proceeding Regional Academic Conference 2009, Penang, Malaysia, 16–17 Dec 2009
8. W.Y. Wan Ahmad, M.R. Ahmad, M.I.A. Kadir, N.A. Zulkifli, S. Misran, R.A. Halim, W.A. Ahmad, Z.A. Zakaria, Colorant from bacteria. The Textile Institute Centenary Conference 2010 Proceedings, Manchester, 3–4 Nov 2010
9. W.A. Ahmad, W.Y. Wan Ahmad, Z.A. Zakaria, N.Z. Yusof, *Application of Bacterial Pigments as Colorant. The Malaysian Perspective*. Springer Briefs in Molecular Science (Springer, Heidelberg, 2012). doi:10.1007/978-3-642-24520-6. ISBN 978-3-642-24519-0
10. A. Shirata, T. Tsukamoto, H. Yasui, T. Hata, S. Hayasaka, A. Kajima, H. Kato, Isolation of bacteria producing bluish-purple pigment and use for dyeing. Jpn. Agric. Res. Q. **34**, 131–140 (2000)
11. M. Moss, Bacterial pigments. Microbiologist, 10–12 (2002)
12. J. Vaidyanathan, Z. Bhathena-Langdana, R. Adivarekar, M. Nerurkar, Production, partial characterization, and use of a red biochrome produced by *Serratia sakuensis* subsp. nov strain KRED for dyeing natural fibers. Appl. Biochem. Biotechnol. **166** (2), 321–335 (2012)
13. R. Adivarekar, M. Nerurkar, J. Vaidyanathan, Z. Bhathena-Langdana, *Analysis of Dyeing Properties of a Bluish Violet Pigment Extracted from Chromobacterium spp.* (International Dyer, World Textile Publications, 2011)
14. D.S. Diana, M. Moresi, A.M. Gallo, M. Petruccioli, Assessment of dyeing properties of pigments from *Monascus purpureus*. J. Chem. Technol. Biotechnol. **80**, 1072–1079 (2005)
15. F.A. Nagia, R.S.R. EL Mohamedy, Dyeing of wool with natural Anthraquinone dyes from *Fusarium oxysporum*. Dyes Pigments **75**, 550–555 (2007)
16. M. Nerurkara, J. Vaidyanathanb, R. Adivarekara, Z.B. Langdanab, Use of a natural dye from *Serratia marcescens* subspecies *Marcescens* in dyeing of textile fabrics. Octa J. Environ. Res. **1**(2), 129–135 (2013)
17. C.K. Venil, Z.A. Zakaria, W.A. Ahmad, Bacterial pigments and their applications. Process Biochem. **48**, 1065–1079 (2013)
18. Y. Lu, L. Wang, Y. Xue, C. Zhang, X.H. Xing, K. Lou, Z. Zhang, Y. Li, G. Zhang, J. Bi, Z. Su, Production of violet pigment by a newly isolated psychrotrophic bacterium from a glacier in Xinjiang, China. Biochem. Eng. J. **43**, 135–141 (2009)
19. N. Sundaramoorthy, P. Yogesh, R. Dhandapani, Production of prodigiosin from *Serratia marcescens* isolated from soil. Ind. J. Sci. Technol. **2**(10) (2009)

FSC
www.fsc.org
MIX
Papier | Fördert gute Waldnutzung
FSC® C083411

Kolibri 360 GmbH
Ferdinand-Jühlke-Straße 7
99095 Erfurt, Deutschland
produktsicherheit@kolibri360.de